TRENDS IN RESEARCH ON MICROSTRIP ANTENNAS

Edited by **Sudipta Chattopadhyay**

Trends in Research on Microstrip Antennas
http://dx.doi.org/10.5772/65580
Edited by Sudipta Chattopadhyay

Contributors

Mohammad Alibakhshikenari, Ernesto Limiti, Bal Singh Virdee, Vivek Singh Kushwah, Geetam Singh Tomar, Seyed Ali Razavi Parizi, Milind Thomas Themalil, Alexandre Jean René Serres, Paulo Fernandes Silva Júnior, Paulo Silva, Georgina Karla Freitas Serres, Raimundo Carlos Silvério Freire, Tulio Chaves Albuquerque, Josiel Cruz, Maciel Oliveira, Kamel Sultan

Published by InTech

Janeza Trdine 9, 51000 Rijeka, Croatia

Notice

Statements and opinions expressed in the chapters are these of the individual contributors and not necessarily those of the editors or publisher. No responsibility is accepted for the accuracy of information contained in the published chapters. The publisher assumes no responsibility for any damage or injury to persons or property arising out of the use of any materials, instructions, methods or ideas contained in the book.

Publishing Process Manager Romina Skomersic
Technical Editor SPi Global
Cover InTech Design team

First published November, 2017
Printed in Croatia
Legal deposit, Croatia: National and University Library in Zagreb

Additional hard copies can be obtained from orders@intechopen.com

Trends in Research on Microstrip Antennas, Edited by Sudipta Chattopadhyay
p. cm.
Print ISBN 978-953-51-3601-9
Online ISBN 978-953-51-3602-6

For EU product safety concerns:
IN TECH d.o.o., Prolaz Marije Krucifikse Kozulić 3, 51000 Rijeka, Croatia, info@intechopen.com or visit our website at intechopen.com.

Meet the editor

Dr. Sudipta Chattopadhyay received his BSc degree (Physics Honours) from the University of Calcutta and did BTech, MTech, and PhD degrees from the Institute of Radio Physics and Electronics, University of Calcutta in 1999, 2001, and 2011, respectively. Since then, he has started his independent research in the field of Antenna Engineering. He is currently working as an associate professor and head at the Department of Electronics and Communication Engineering, Mizoram University (a Central University, Government of India), Mizoram, India. Before joining Mizoram University, he served Siliguri Institute of Technology, West Bengal, India, for 15 years as a faculty member. His area of research includes microwave antennas, microstrip and integrated antennas, defected ground structures, and computer-aided design of patch antennas. He regularly serves as the reviewer of *IEEE Antennas and Propagation Magazine*; *IEEE Antennas and Wireless Propagation Letters*; *IET Microwaves, Antennas, & Propagation* journal, UK; *IEEE Transactions on Antennas and Propagation*; *International Journal of RF and Microwave Computer-Aided Engineering*, Wiley; *International Journal of Microwave and Wireless Technologies*, Cambridge; and also *Taylor and Francis* journal. He has more than 60 publications in referred international journals and international conferences. He has contributed several chapters in different edited research handbooks and is presently acting as a sole editor in the handbook of *Trends in Research on Microstrip Antenna*. His research is cited in different research handbooks as well as in well-known undergraduate text book of *Antennas and Wave Propagation* by J. D. Kraus et al. fourth edition, Tata McGraw-Hill Publishing Co. Ltd. In the last 2 years, he has been working as the editorial board member of Wiley journal *International Journal of RF and Microwave Computer-Aided Engineering* and *Journal of Electromagnetic Analysis and Applications*. He is listed in *Marquis Who's Who in the World*, USA, 26th Edition, 2009 and also listed in *2000 Outstanding Intellectuals of the 21st Century*, UK, 2010.

Contents

Preface

In the present era, with the expeditious development of modern wireless communication systems, unidirectional antennas having low profile, low cross polarization, and wide beam radiation patterns with stable gain are in the spotlight in the current research on the subject. In this context, microstrip patch antennas are becoming increasingly popular as they have small volume and a low-profile planar configuration with low fabrication cost. Therefore, microstrip patch antenna has become a major focus to the researchers in the field of antenna engineering.

The fundamental geometry of a microstrip patch antenna consists of a metallic patch printed on a grounded substrate. Although the concept of microstrip patch as radiator was proposed in the early 1950s, still till late 1970s, this type of antenna did not attract serious attention of the antenna community.During the last three decades, the scientists and researches have put significant effort to establish such tiny radiator with significantly improved input and radiation performances.

The progress in modern tiny multifunctional wireless devices has dramatically increased the demand for such antennas in recent years. Furthermore, in the last few years, such microstrip antennas found numerous applications in both the military and the commercial sectors.

In this book, some recent advances in microstrip antennas are presented. This book contains mainly three sections. In the first section, some new approaches to modern analytical techniques rather than the conventional cavity model, transmission line model, or spectral domain analysis have been discussed. Although the microstrip patch antenna is a very promising candidate in modern wireless scenario, still it suffers from severe limitation of poor bandwidth performance. Hence, the investigation for improving its bandwidth performance is a challenging task to antenna researchers. In the second section of the book, a light has been showered on some new techniques for bandwidth enhancement of microstrip radiators. In the last section of the book, the recent trends in microstrip antenna research have been showcased. Some newfangled application-oriented approach to this field is vividly discussed.

The book's main objective is to facilitate the microstrip antenna researchers for exploring the subject in more vibrant manner and also to revolutionize wireless communications. A sufficient number of topics have been covered, some for the first time in a research handbook. I hope that the book will surely be beneficial for scientists, practicing engineers, and researchers working in the field of microstrip antennas.

I would like to acknowledge the authors of this book for their invaluable contributions. It is my pleasure to acknowledge the precious suggestions and constructive criticisms received from Prof. Debatosh Guha, fellow, IEEE; Dr. Inder Bahl, editor in chief, *International Journal of RF and Microwave Computer-Aided Engineering*, Wiley; Prof. L. Lolit Kumar Singh, dean,

School of Engineering and Technology, Mizoram University; Dr. J. Y. Siddiqui of Calcutta University; Mr. A. Ghosh of Mizoram University; and Dr. Ross Stone, past editor of *IEEE Antennas and Propagation Magazine*.

I am highly indebted to my scholars Major Umesh Ankush Pawar, scientist, DRDO, and Mr. Subhradeep Chakraborty, scientist, CEERI, Pilani, for their invaluable suggestions and continuous support during this book project.

It is my gratification to offer thanks to Ms. Romina Skomersic, publishing process manager, InTechOpen publisher for her continuous help without which the book project would not be successful. I am also thankful to all the editorial and technical members associated with this book project for their support.

Finally, I must pay tribute to my parents (Salil Kumar Chatterjee and Rekha Chatterjee) and my wife Mrs. Moumita Chattopadhyay for their support, patience, and sacrifice during the work on this book.

Dr. Sudipta Chattopadhyay
Department of Electronics and Communication Engineering
Mizoram University (a Central University, Government of India)
Aizawl, Mizoram, India

Section 1

Microstrip Antennas Bandwidth

Chapter 1

Bandwidth Enhancement Techniques

Seyed Ali Razavi Parizi

Additional information is available at the end of the chapter

http://dx.doi.org/10.5772/intechopen.70173

Abstract

In this chapter, a variety of procedures proposed in the literature to increase the impedance bandwidth of microstrip patch antennas are presented and discussed. Intrinsic techniques, proximity coupled and aperture-coupled patches, applying horizontally coupled patches to driven patch on a single layer and stacked patches are discussed. Beside the linear polarised solutions, some techniques for designing wideband circular polarised patch antennas are also presented. Furthermore, some other techniques proposed in the literature including log-periodic array of patches, E-shaped patch, L-shaped feeding, microstrip monopole slotted antenna, defected ground/patch technique and the latest works during the recent years are introduced and investigated. It is tried to make a comparison between different methods giving a typical bandwidth that can be obtained using each method, beside discussing about the benefits or limitations that each method has.

Keywords: bandwidth enhancement, microstrip patch antenna, aperture coupled, stacked patches, non-contact feeding, parasitically coupled, sequential rotated array

1. Introduction

Microstrip patch antennas, in their conventional form, are narrow-band structures. Their impedance bandwidth is typically 1–2%. This can be attributed to two factors: the resonant style of antenna (which makes the antenna radiate efficiently only over a narrow band of frequencies) and the thin thickness of antenna, typically less than 0.05λ0.

This feature of conventional microstrip patch antennas makes them unsuitable in many applications where quite wide bandwidth is required. So, many researches have been done during the past few decades to overcome this limitation and several procedures have been proposed.

In this chapter, we will describe the most useful procedures proposed in the literature to increase the impedance bandwidth of microstrip patch antennas. In Section 2, the intrinsic techniques which can be used to increase the bandwidth of a single layer direct fed microstrip patch antenna are described. Section 3 investigates different feeding techniques used for bandwidth enhancement of microstrip patch antennas. In the proposed techniques, the antenna structure is multilayered and the patch is fed by a non-contact feed network. In Section 4, bandwidth enhancement by use of horizontally parasitic elements at the antenna aperture will be described. Sections 5 and 6 explore the stacked patches, a multilayered solution using vertically parasitic elements that can result in very wide bandwidths. In Sections 2–6, the focus is on linear polarised antennas. So, the solutions for wideband circular polarised patch antennas are introduced and discussed in Section 7. Finally, in Section 8, other techniques proposed in the literature including log-periodic array of patches, E-shaped patch, L-shaped feeding and microstrip monopole slotted antenna are introduced and investigated.

In this chapter, we tried to discuss about the principles of operation of presented antennas to give the reader insight into how these antennas work. Some examples are also given. The advantages and limitations of each method will be described and a comparison between them will be provided by giving the typical order of bandwidth that can be achieved using each method. By this way, in this chapter, it is tried to provide a designer's prospective of different techniques used for bandwidth enhancement of microstrip patch antennas.

2. Intrinsic techniques

Two intrinsic procedures can be applied to improve the bandwidth of a single layer direct fed microstrip patch antenna. One is increasing the substrate thickness and the other is decreasing the dielectric constant of antenna substrate (relative permittivity of near to one). This can also be observed in **Figure 1**, in which the bandwidth trends of a direct fed single layer microstrip patch relative to substrate permittivity and thickness are shown.

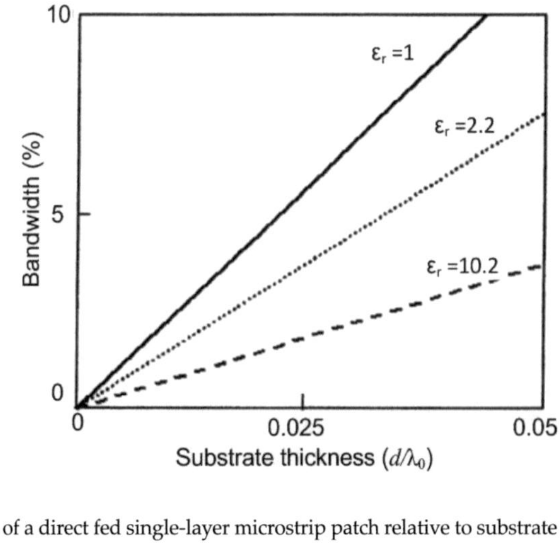

Figure 1. Bandwidth trends of a direct fed single-layer microstrip patch relative to substrate permittivity and thickness [1].

http://dx.doi.org/10.5772/intechopen.70173

In fact, by increasing the thickness of the radiating patch substrate, the aperture size through which the fields are radiating to the space can be enhanced leading to easier impedance matching of antenna aperture which results in antenna bandwidth enhancement. On the other hand, by decreasing the substrate permittivity (making it closer to the permittivity of outer space), the reflection coefficient at the antenna aperture is reduced making it easier to impedance match which provides wider bandwidth for the antenna.

It should be noted that as the thickness of the substrate is increased and the dielectric constant is reduced, the patch size becomes smaller which provides lower gain. In addition, the radiated power by feed network increases leading to more spurious radiation and higher cross-polarisation level. Furthermore, the surface waves are more excited reducing the antenna radiation efficiency.

Taking the abovementioned limiting factors into consideration, besides the fact that there are limited values for thicknesses and dielectric constants provided by standard commercial substrates, the bandwidth enhancement by the mentioned intrinsic techniques cannot exceed 10% [2] which is still not adequate for many applications like L-band radar which needs 19% bandwidth (1.4–1.7 GHz) or C-band satellite TV that requires 12.5% bandwidth (3.7–4.2 GHz). So, other techniques as mentioned at the following must be applied to enhance the bandwidth more.

3. Feeding techniques

In all proposed feeding techniques presented in this section, the antenna structure is multi-layered and the radiating patch is fed by a non-contact feed network. In fact by introducing a coupling mechanism between the feed network and patch, a resonance is created at the vicinity of patch resonance which can result in antenna bandwidth enhancement if the feed network and patch are well coupled.

3.1. Proximity coupled feed

In **Figure 2**, topology of proximity coupled feeding is depicted. In this feeding mechanism, the radiating patch on the upper layer (radiating layer) is excited by an open-ended microstrip feed line printed on the lower layer (feed layer). It can be seen that there is no direct contact

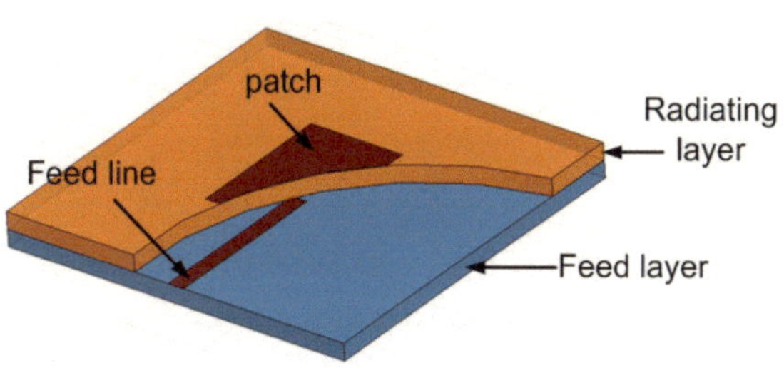

Figure 2. Topology of proximity coupled feed.

between feed line and radiating patch. So, the fields are coupled to the patch via the open end of stub. By tuning the location of open end of feed line relative to the patch, a proper coupling between them can be obtained leading to expansion of antenna bandwidth. By this way, the bandwidth at the order of 8% can be achieved.

In order to increase the bandwidth more, we can increase the thickness of radiating layer. However, this makes the coupling of power from the feed network to the patch more difficult. To solve this problem, thicker substrate for the feed layer can be used. In this case, more power is coupled to the patch via open circuit stub, but at the same time, the spurious radiation caused by feed network becomes more severe which can degrade the antenna efficiency. Another solution for this issue is applying a simple matching structure, for example, a quarter-wave transformer, to the feed line. By this way, i.e. using thicker laminate for the radiating layer besides applying matching circuit to the feed network, the bandwidth can be doubled [1]. In **Figure 3**, one example about this issue is presented [1]. In this example by doubling the thickness of radiating layer and using a simple impedance transformer at the feed line, the bandwidth could be enhanced from 8 to 14%.

3.2. Aperture-coupled feed

In **Figure 4**, the topology of aperture-coupled feeding is depicted. In this feeding mechanism, the radiating patch on the upper layer (radiating layer) is excited by a microstrip feed line printed on the lower layer (feed layer) via an aperture (slot) etched at the ground plane of radiating layer.

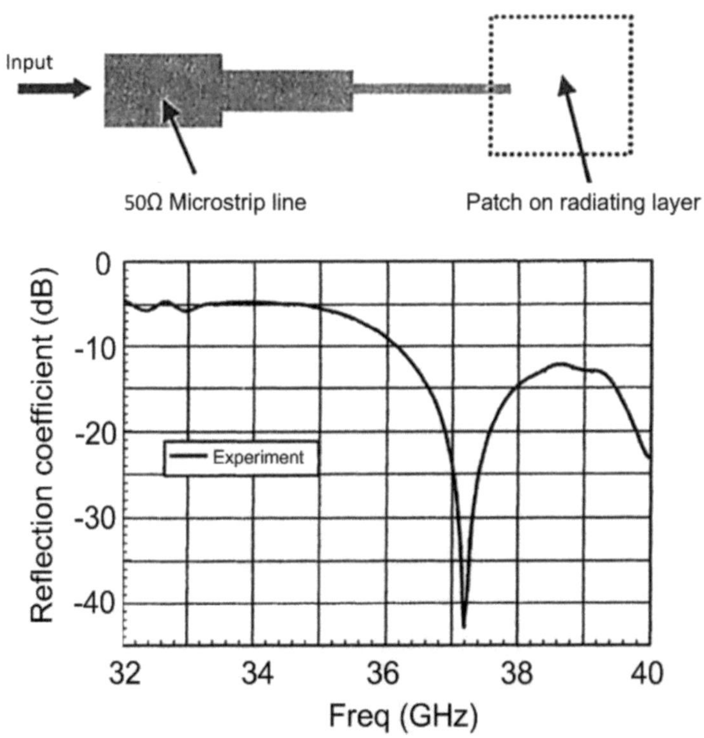

Figure 3. S11 of a proximity coupled microstrip patch antenna presented in Ref. [1].

http://dx.doi.org/10.5772/intechopen.70173

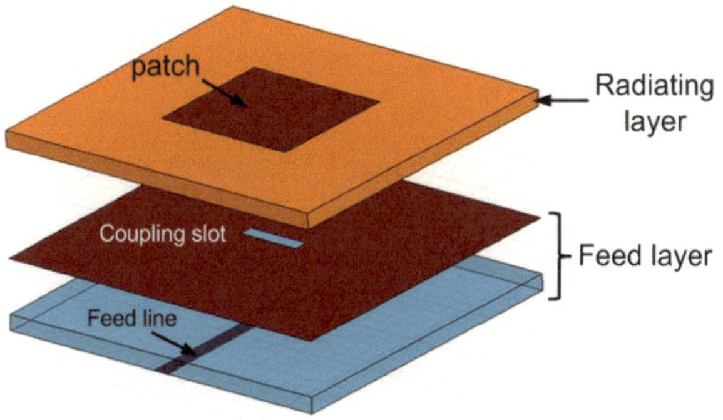

Figure 4. 3D distributed view of aperture-coupled fed microstrip patch antenna.

In this structure, two resonances are provided, one by the patch and the other one by slot. If these two radiators are properly coupled, then the two corresponding resonances become close to each other leading to antenna bandwidth enhancement. The proper coupling between slot and patch can be obtained by tuning the dimensions of slot. By this way, the bandwidth at the order of 20–30% can be obtained. In Ref. [1], a sample is designed for mobile communication base station application with operation frequency of 1.9 GHz, bandwidth of 27% and 9 dBi gain.

In order to increase the bandwidth more, thick dielectric can be used for radiating layer. In this case, the slot size should be increased to make sure that the power is still coupled to the patch properly. In Ref. [3], 40% bandwidth has been reported using this simple technique.

In this technique, the shape and the size of the coupling slot can considerably affect the coupled power and consequently the antenna bandwidth and efficiency. In general, as the coupling power gets stronger by changing the slot shape, thicker substrate for the radiating layer can be used leading to wider impedance bandwidth. On the other hand, as the slot size decreases, the backward radiation which is dominantly caused by the slot decreases enhancing the antenna efficiency. In fact, by choosing the proposer shape for the slot, we try to provide the maximum

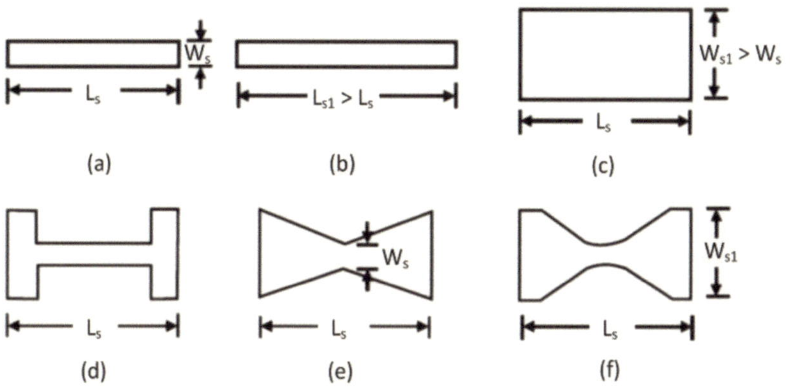

Figure 5. Different shapes used for the coupling slot: (a). thin rectangular, (b). longer rectangular, (c). wider rectangular (d). 'H'-shaped, (e). Bowtie-shaped and (f). hourglass-shaped [2, 4].

coupling with the smallest size of the slot. By this way, we can more improve the bandwidth by increasing the radiating layer thickness or reducing its dielectric constant while the maximum efficiency is guaranteed.

One of the most used shapes for the coupling slot is a thin rectangle (**Figure 5(a)**) by which strong coupling can be given with a simple design. However, stronger coupling can achieved by making it longer or wider (as shown in **Figure 5(b** and **c)**). The 'H'-shaped and bowtie-shaped slots (**Figure 5(d** and **e)**) can provide stronger coupling than the rectangular slot since the fields magnitude in rectangular slot has sinusoidal variation; however, in 'H'-shaped and bowtie-shaped slots, it is quite uniform. By combining the 'H'-shaped and bowtie-shaped slots, i.e. making an hourglass-shaped slot shown in **Figure 5(f)**, even more uniform field distribution across the slot and consequently stronger coupling can be achieved [2, 4].

4. Parasitic elements on the single layer

By introducing some dummy elements properly coupled to the driven element at the same radiating layer, i.e. by introducing parasitic patches horizontally coupled to the driven patch, the overall bandwidth of the antenna can be enhanced if the resonant frequency of the coupled elements are slightly different to that of the driven patch since, as shown in **Figure 6**, the overall frequency response is the superposition of the frequency responses of individual patches. In fact, in this technique, both the driven patch and parasitic elements are placed at

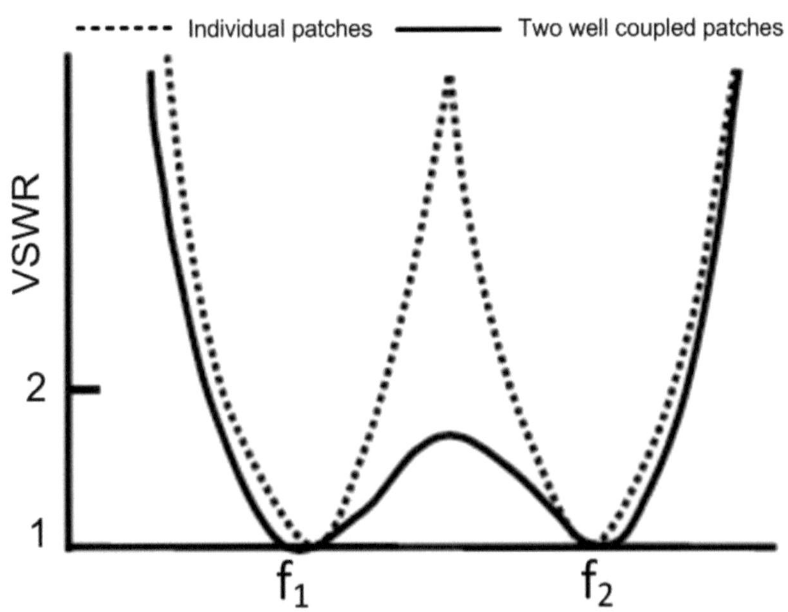

Figure 6. Bandwidth enhancement by introducing a dummy patch coupled to the driven patch while the resonance frequencies of the two resonators are quite different [2].

http://dx.doi.org/10.5772/intechopen.70173

the same layer, meaning that this method is a single layer solution providing easy process for antenna fabrication. The main shortcoming of this method is that the antenna overall size is large leading to the problem of grating lobes for the case, an array is going to develop. In addition, the asymmetry of antenna structure with respect to the centre of driven patch, tends to degrade the radiation pattern. This technique has been realised in four different ways that are described in the following.

4.1. Coupling via radiating edges

In **Figure 7(a)**, a microstrip patch antenna using two unequal parasitic elements at the radiation layer is presented [2, 5]. In the proposed antenna, the driven patch is fed by a probe and two dummy patches are placed at its both sides. In this structure, the parasitic patches are coupled to driven patch via the radiating edges of driven patch. The dimensions of patches are quite different providing three slightly different resonances. The coupling between patches can be tuned by the gaps between them controlling the impedance matching of the antenna.

By this technique, the bandwidth at the order of 20% can be obtained [2]. In **Figure 7(b)**, the frequency response of a sample design presented in Ref. [5] is shown. In this figure, three close well-coupled resonances can be observed which have resulted in antenna bandwidth enhancement. The resulted bandwidth relative to the centre frequency is 10.6% (3.1–3.45 GHz).

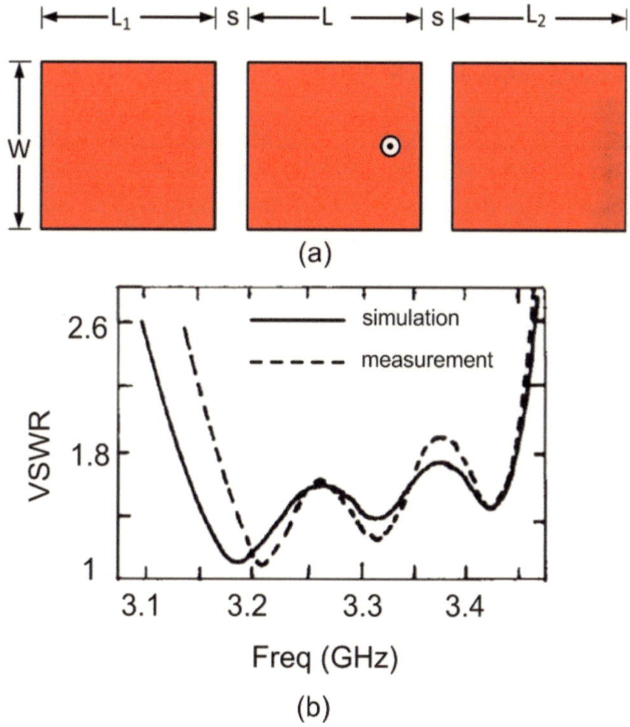

Figure 7. Microstrip patch antenna coupled to two parasitic elements at the radiation layer via the radiating edges (a). Antenna topology (b). VSWR plot for a sample designed in Ref. [5].

4.2. Coupling via non-radiating edges

In this method, in the same way as shown in **Figure 7**, two unequal parasitic elements are placed at both sides of the driven patch. But here, as shown in **Figure 8(a)**, the parasitic patches are coupled to driven patch via the non-radiating edges of the driven patch. Since the fields are not uniform at non-radiating edges, unlike at radiating edges where the fields are uniform, the coupling between driven patch and parasitic elements is weaker compared to the previous case where the coupling is provided via radiating edges. So here the gaps between the driven patch and parasitic elements should be smaller than those in the previous case.

By this technique, the bandwidth at the order of 20% can be obtained. In **Figure 8(b)**, the frequency response of a sample design presented in Ref. [2] is shown. In this figure, also three close well coupled resonances can be observed which has resulted in antenna bandwidth enhancement. The resulted bandwidth relative to the centre frequency is 14.5% (2.8–3.3 GHz).

4.3. Coupling via four edges

By mixing the two previous procedures, i.e. placing four patches around the driven patch coupled to it via both radiating and non-radiating edges (as shown in **Figure 9(a)**), further bandwidth enhancement and more gain can be achieved. In Ref. [6], 18.2% bandwidth is obtained using this method. In **Figure 9(b)**, the frequency response of the sample presented in Ref. [6] is shown.

4.4. Annular ring loaded shorted patch

In **Figure 10**, a circular shorted microstrip patch loaded by an annular ring is depicted. In this structure, the driven element is the circular patch which is shorted to the ground plane and the parasitic element is the annular ring surrounding it. The dimensions of both the shorted patch and the ring should be well-determined so that strong coupling between them is formed. Then, by increasing the spacing between them, we can reduce the coupling making it possible to increase the bandwidth. In Ref. [7], a sample of this antenna is designed in which 6.6% bandwidth and 8.4 dBi gain are obtained.

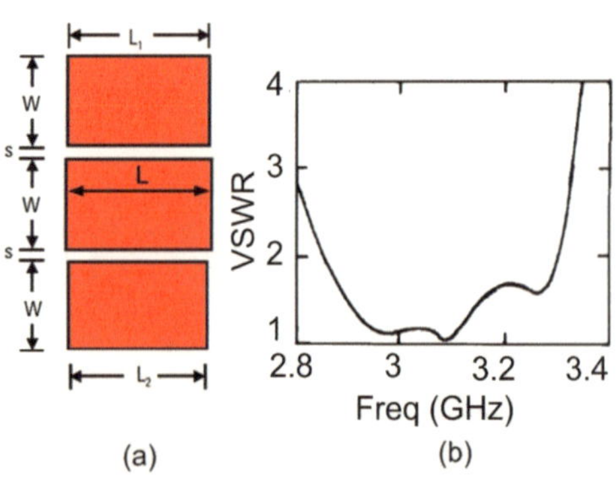

Figure 8. Microstrip patch antenna coupled to two parasitic elements at the radiation layer via the non-radiating edges. (a). The configuration of patches and (b). VSWR plot for a sample of antenna presented in Ref. [2].

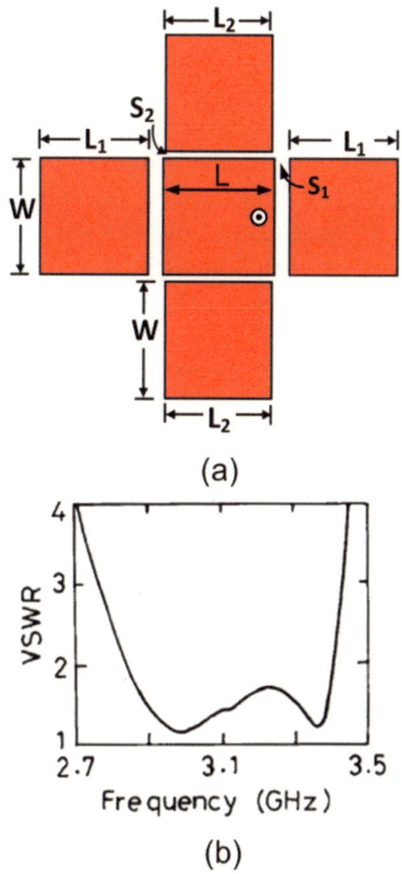

Figure 9. Microstrip patch antenna coupled to four parasitic elements at the radiation layer via both radiating and non-radiating edges. (a). The configuration of patches and (b). VSWR plot for a sample of antenna presented in Ref. [6].

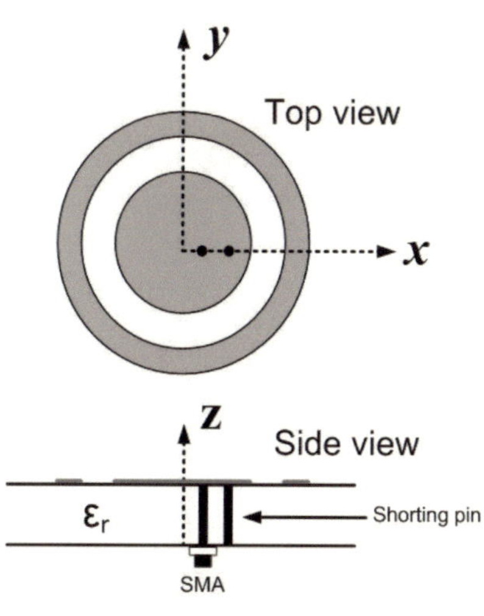

Figure 10. Topology of an annular ring loaded shorted patch [7].

5. Stacked patches

5.1. Configuration

In microstrip patch antennas, the term 'stacking' is used for the case where the driven patch is vertically coupled to another patch. By this solution, we do not have the problem of antenna large planar size (which was the issue for the solutions presented in the previous section in which the coupling between parasitic patches and the driven patch were established horizontally) solving the problem of grating lobes when the antenna is used in an array. However, in this technique, since the antenna structure is dual layered, the fabrication process is quite harder than the single-layered antennas presented in previous section. Using this technique, the bandwidth at the order of 20% can be obtained.

In **Figure 11**, a typical probe-fed rectangular stacked patch antenna is shown. It is a dual-layer structure in which the driven patch is located at the lower layer and coupled to the parasitic patch printed on the upper layer. In this configuration, typically rectangular patches are used, however, circular and annular patches can also be used in the same manner. Annular ring (by inner radius) and rectangular (by width) stacked patches have extra degree of freedom compared to circular stacked patches which gives them easier impedance matching control at the expense of slightly reduced bandwidth [1]. In **Figure 12**, the VSWR plot of a sample antenna designed in Ref. [2] using this technique is illustrated. We can see that the bandwidth (VSWR < 2) of 18.8% is obtained.

In the configuration shown in **Figure 11**, there are two resonators, one is the driven patch and the other is the parasitic patch. In design process, the dimensions of patches are chosen so that they resonate at the same frequency. Since two different substrates with different dielectric constants (εr_1, εr_2) are used, the dimensions of the two patches are different. The mutual coupling between patches shifts the two resonances with respect to each other producing a mutual resonance making it possible to increase the bandwidth. By proper choosing the substrates' thicknesses and permittivities, the proper coupling between the two resonances

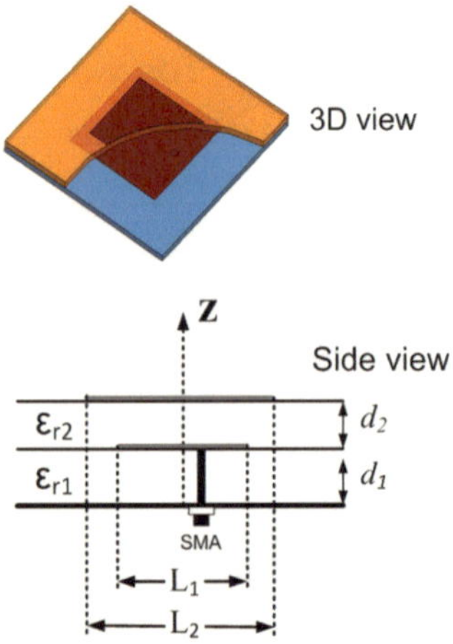

Figure 11. The concept of stacking. A rectangular stacked patch.

can be achieved leading to bandwidth enhancement. Broader impedance bandwidth can be obtained by lowering the coupling using high permittivity, say 10, for the lower substrate and low permittivity, say 1, for the upper one [1]. By this way, high surface wave efficiency is also achieved giving the antenna, a better radiation performance.

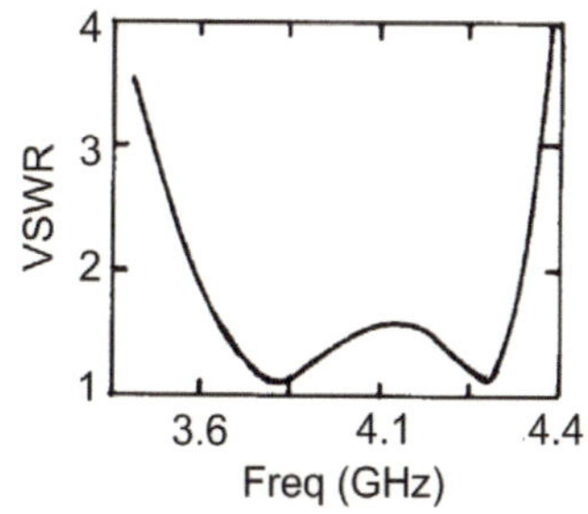

Figure 12. VSWR plot for a sample of antenna presented in **Figure 11** [2].

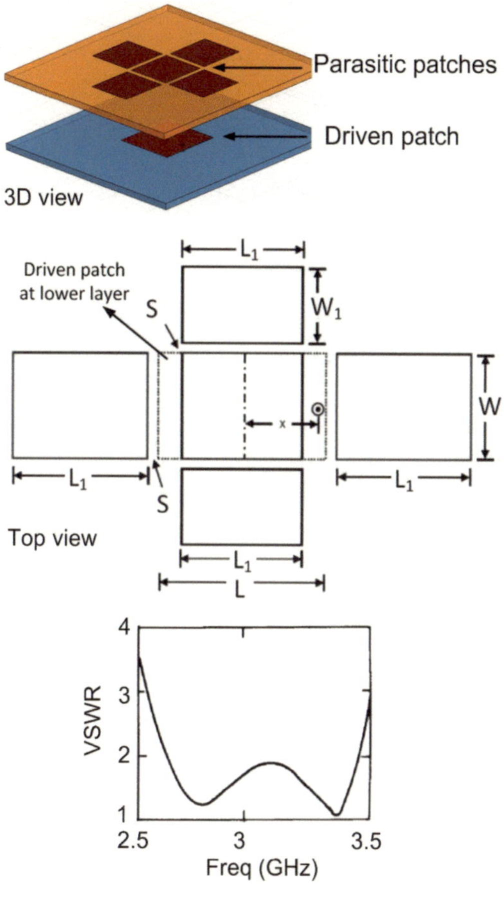

Figure 13. The configuration and frequency response of antenna presented in Ref. [2]. Use of stacking concept for the parasitic elements technique presented in Section 4.

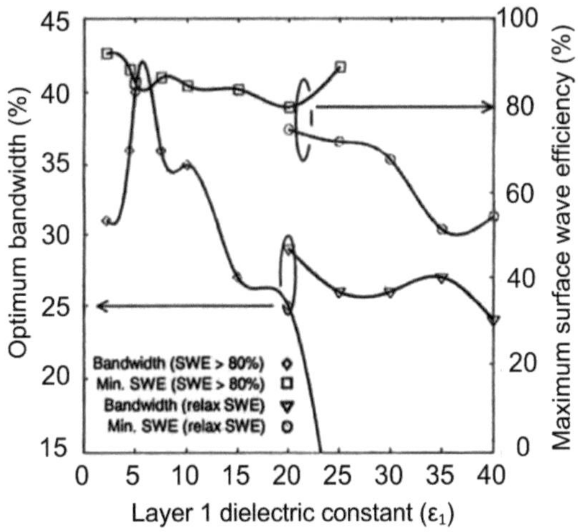

Figure 14. Impedance bandwidth and surface wave efficiency (SWE) vs. the lower layer dielectric constant (ε_{r1}). The relative permittivity of upper layer is 1.07 [1].

The concept of stacking can also be used for the parasitic elements technique presented in Section 4, placing the driven patch on the lower layer, while the parasitic patches are placed on the upper layer. In Ref. [2], 30% bandwidth could be obtained using one rectangular patch on the bottom layer as the driven path and five parasitic rectangular patches placed on the top layer. **Figure 13** shows the configuration and its frequency response.

5.2. Rules of thumb for substrate permittivity

In **Figure 14** [1], the widest impedance bandwidth and minimum surface wave efficiency (SWE) that can be obtained using different laminates for the lower layer while the upper layer uses foam with the relative permittivity of 1.07 as the substrate, are illustrated. It should be noted that the relative permittivity for the upper layer is chosen close to 1 in order to guarantee the maximum surface wave efficiency and bandwidth obtained by tuning ε_{r1}. We can see that with $\varepsilon_{r1} < 15$, wide impedance bandwidths (more than 25%) and high efficiency (more than 80%) can be obtained. However, for $\varepsilon_{r1} > 15$, the surface wave efficiency starts to reduce degrading the radiation performance of the antenna. The reason is that as ε_{r1} is lower, the surface wave associated with the lower patch is more effectively coupled to the upper patch and consequently to the radiating fields improving the antenna efficiency. As a result, higher ε_{r1} leads to lower surface wave efficiency.

6. Aperture-stacked patches

The aperture-stacked patches are multilayered structures by which the bandwidths at the order of 50–70% can be achieved. **Figure 15** shows the general configuration of an aperture-stacked patch antenna. It consists of N dielectric layers, two patches and one ground plane

http://dx.doi.org/10.5772/intechopen.70173

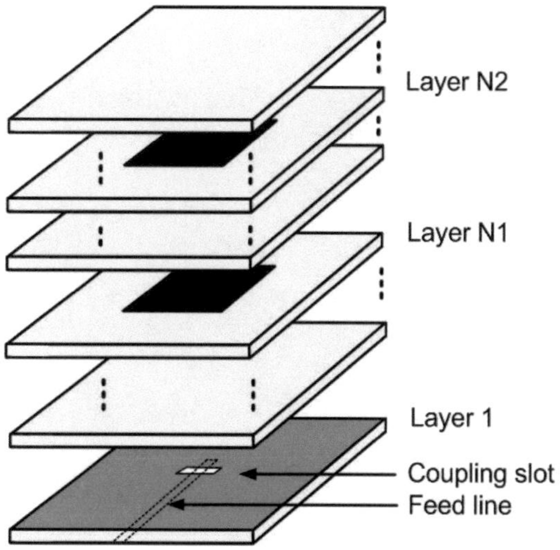

Figure 15. The general configuration of an aperture-stacked patch antenna [1].

at which the coupling aperture, say the slot, is etched. In this configuration, in contrast with the ordinary stacked patches shown in **Figure 11**, the driven patch (the lower one) is fed indirectly introducing one more coupling mechanism (between feed line and driven patch) which results in wider impedance bandwidth.

In **Figure 16**, the reflection coefficient, S11, of a triple layer aperture-stacked patch antenna designed in Ref. [1] for Ka band is presented. It can be observed that the impedance bandwidth of more than 46% relative to centre frequency is obtained. The designed antenna gain is greater than 6 dBi across 26–40 GHz. It should be noted that the presented stacked patch antenna in Ref. [1] is backed by a cavity to increase both front to back ratio and gain.

In **Figure 17**, an aperture-stacked patch antenna with its geometrical dimensions is shown. In this structure, two offset feed lines are used to excite the slot in order to provide more control

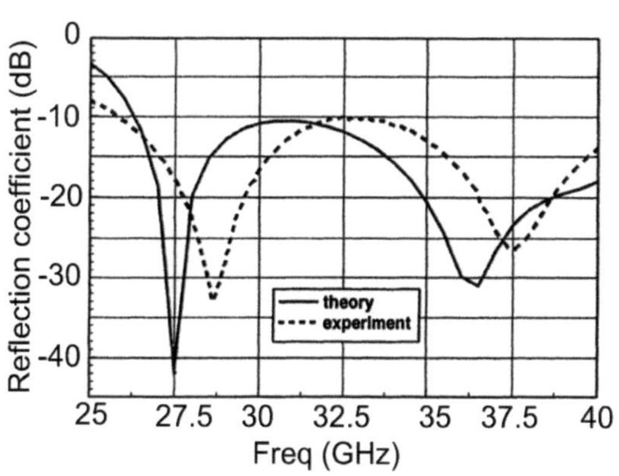

Figure 16. Reflection coefficient of aperture-stacked patch antenna presented in Ref. [1].

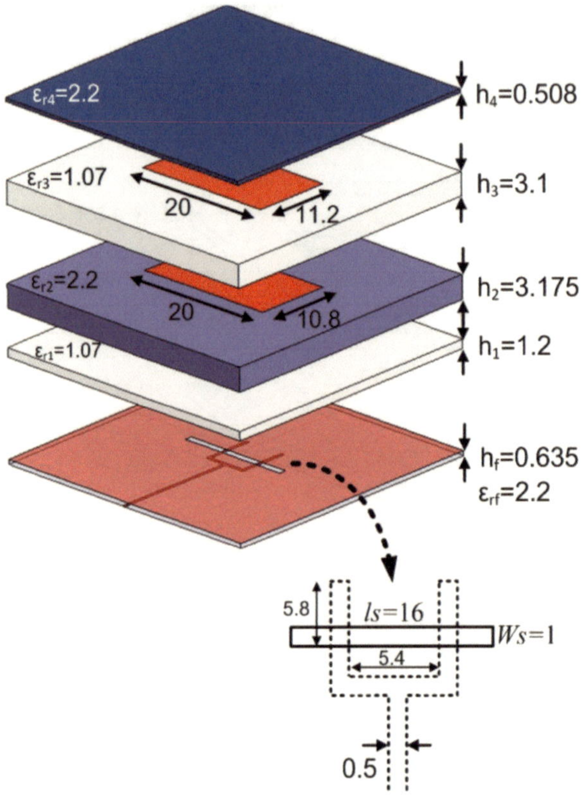

Figure 17. The configuration and dimensions of aperture-stacked patch antenna presented in [1, 8].

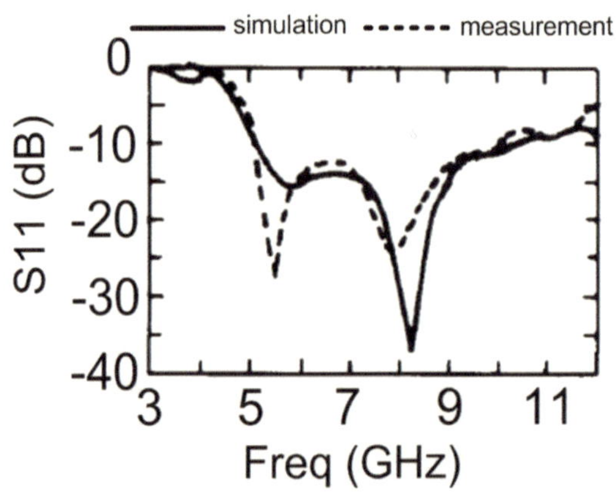

Figure 18. Reflection coefficient of aperture-stacked patch antenna presented in [1, 8].

on coupling power. The two feed lines are 100 Ω lines connected to a 50 Ω line by a power divider. The reflection coefficient, S11, of the designed antenna is shown in **Figure 18** [1, 8]. It can be observed that the impedance bandwidth of about 67% relative to centre frequency is obtained. The designed antenna gain is about 7 dBi over its operating bandwidth.

http://dx.doi.org/10.5772/intechopen.70173

7. Broadband techniques for circular polarisation

In previous sections, the broadband techniques for linear polarisation were discussed. However, in many applications such as radar and navigation systems, circular polarisation is required [2]. The axial ratio (AR) bandwidth of conventional microstrip patch antennas is at the order of 1% which is very narrow for many applications. Therefore, some techniques should be applied to increase it. In this section, several configurations that produce wideband circularly polarised radiation are briefly discussed.

7.1. Parasitic elements on a single layer

This method uses the same concept used in Section 4 for broadening the bandwidth, however, since here the driven patch is circular polarised, the AR bandwidth is increased. In **Figure 19** [9, 10], a dual-fed four edged coupled circular polarised microstrip patch antenna is shown. The driven patch is circularly polarised using two orthogonal ports with 90° phase difference. It is coupled to four parasitic patches around it via its four edges. By this way, the two orthogonal modes of the driven patch are coupled to the two orthogonal series of parasitic patches. As a result, the two orthogonal modes are impedance matched in wider frequency band, compared to the case no parasitic patch is used, leading to AR bandwidth enhancement. Since the two orthogonal modes originated by the driven patch have the same amplitude, the gaps between the driven patch and all four parasitic patches should be kept the same in contrast to **Figure 9(a)** where the gaps sizes in the two orthogonal directions have to be different.

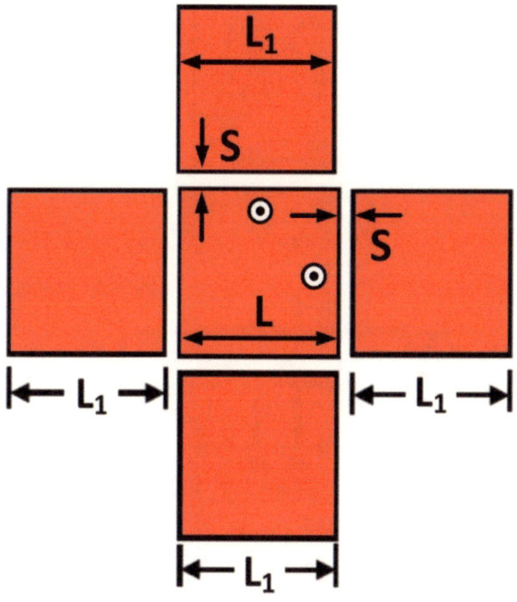

Figure 19. Dual-fed four edged coupled circular polarised microstrip patch antenna [9, 10].

It should be noted that here, in order to keep the amplitude of orthogonal radiating modes well balanced to improve gain, bandwidth and AR, all parasitic patches should be identical keeping the same shapes and the same sizes. Since the driven patch is square shaped, the parasitic patches also should be square shaped as well. By this single-layer technique, the AR bandwidth at the order of 10% can be achieved.

7.2. Stacking

Here, also in the same way as described in Section 5, the AR bandwidth can be enhanced using stacking technique. In **Figure 20**, a circular polarised dual-fed stacked microstrip patch is shown. The driven patch is dual-fed and produces circular polarisation. The two orthogonal modes generated by the driven patch are vertically coupled to the parasitic patch and excite two orthogonal modes inside it. In this case, we can say the two orthogonal modes of lower patch are coupled to the two orthogonal modes of upper one. As a result, these two orthogonal modes are impedance matched over wider frequency band compared to the case no parasitic patch is used leading to wider AR bandwidth. In Ref. [11], a sample antenna base on the configuration shown in **Figure 20** is designed and 18% AR bandwidth is obtained. In Ref. [11], both the driven and parasitic patches are squared.

7.3. Aperture coupling

By changing the slot/patch shape in the aperture coupling technique discussed in Section 3.2, circular polarised radiation with enhanced AR bandwidth can be achieved [12–17]. In **Figure 21**, an aperture-coupled square patch antenna with cross-shaped coupling slot is shown. By this

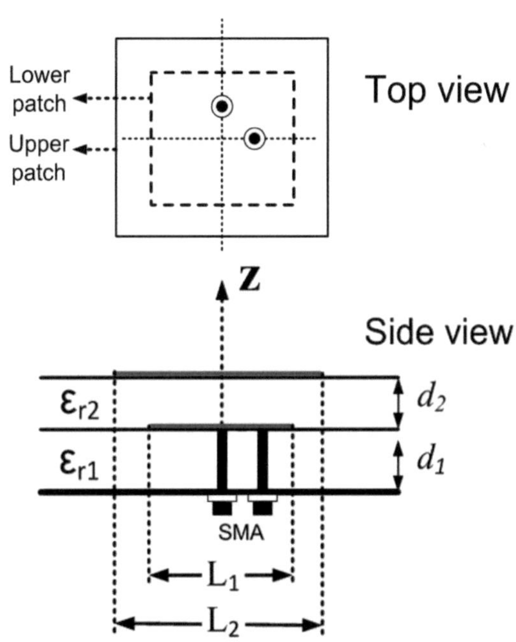

Figure 20. AR bandwidth enhancement by stacking technique.

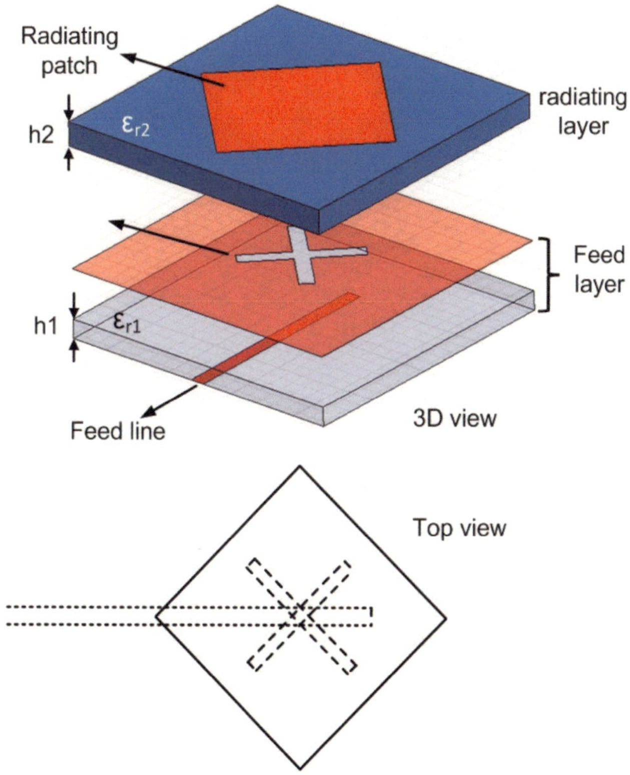

Figure 21. Aperture-coupled microstrip patch antenna using cross-shaped coupling slot. Different lengths are chosen for the two crossed slots to provide 90° phase difference between the two orthogonal modes.

configuration, two orthogonal modes originated by the cross-shaped slot are coupled to the two orthogonal modes of the patch leading to bandwidth enhancement of each mode and consequently expansion of AR bandwidth. In Ref. [12], a sample antenna is designed using the configuration shown in **Figure 21** and 1.4% AR bandwidth is obtained.

In **Figure 21**, two crossed slots and a square patch were used. However, the same performance can be obtained using a single slot and modifying the patch shape. In **Figure 22**, an aperture-coupled patch antenna using a single 45° rotated slot and a nearly square patch are depicted. In Ref. [17], a circular polarised antenna with AR bandwidth of about 1.1% is designed using this configuration.

7.4. Array

Probably the widest AR responses can be obtained using the array technique. In this method, the elements of array are sequentially rotated while fed by equal magnitudes and different phases. The elements are either linear polarised or circular polarised, however, the whole array provides circular polarisation. In **Figures 23** and **24** [18–21], several array configurations for providing wideband circular polarised radiation are shown. In these figures, the array elements are circular polarised.

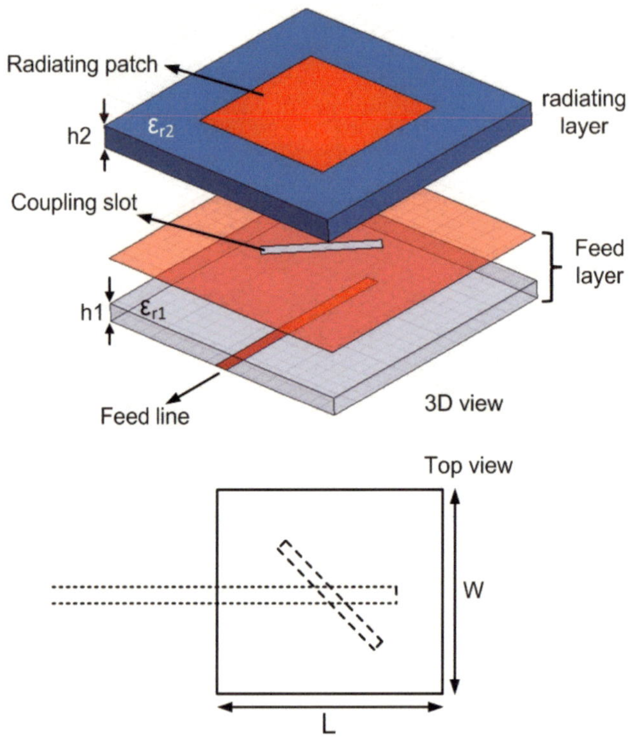

Figure 22. Aperture-coupled patch antenna using a single 45° rotated slot and a nearly square patch.

Figure 23(a) [18] shows a 1 × 2 array in which the elements are 90° out of phase and 90° rotated relative to each other. The 90° phase difference is provided by applying additional length to the feed line. In **Figure 23(b)** [19], a 2 × 2 array using the same concept is shown. In this figure, each element has 90° rotation and 90° phase difference relative to its neighbouring element. In **Figure 24(a)** [20], a 2 × 4 array is illustrated in which the phase of each element is equal to its rotation angle relative to the element fed by 0° phase. A sample of this configuration is designed in

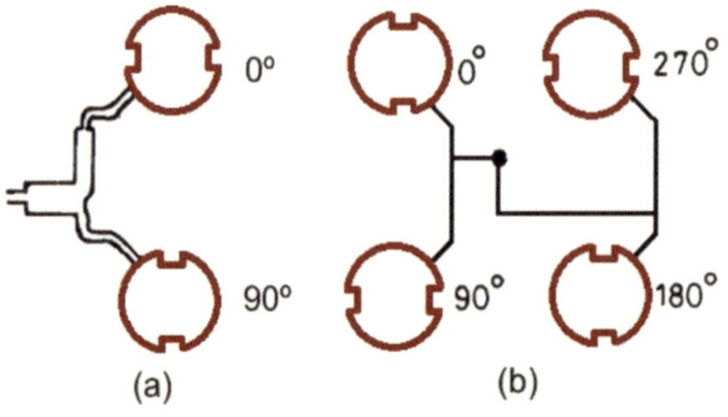

Figure 23. Two sequentially rotated array configurations using circular polarised elements. (a). A 1 × 2 array with 90° rotated and phase-shifted elements. (b). A 2 × 2 array in which each element has 90° rotation and phase difference relative to its neighbouring element [18, 19].

http://dx.doi.org/10.5772/intechopen.70173

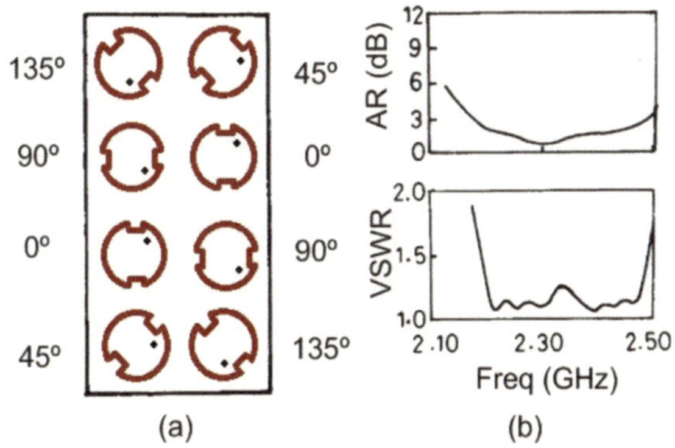

Figure 24. A sample 2 × 4 sequentially rotated array. (a). Antenna configuration and (b). VSWR and AR of designed antenna vs. frequency [20].

Ref. [20] and its frequency responses including VSWR and AR are shown in **Figure 24(b)**. It can be observed that the bandwidth for VSWR < 1.5 is 13.7% and the bandwidth for AR < 3 is 14%.

In the configurations shown in **Figures 23** and **24**, the array elements are in direct contact with feed network. However, non-contact feeding techniques such as proximity coupled and aperture-coupled feedings can also be applied here [21].

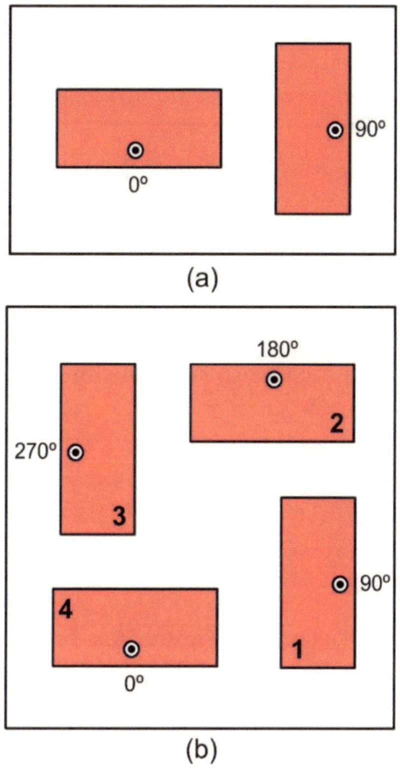

Figure 25. Two sample sequentially rotated arrays using linear polarised elements: (a). a 1 × 2 array and (b) a 2 × 2 array [22].

As mentioned before, wideband circular polarisation using sequentially rotated array can also be achieved using linear polarised elements. In **Figure 25**, two sample arrays are shown [22]. The arrangement of elements is exactly the same as that shown in **Figure 23**. **Figure 25(a)** shows a 1 × 2 array with two linear polarised orthogonal patches fed by 90° phase difference. It is easy to understand that by this configuration, two orthogonal modes with 90° phase difference are radiated by the two patches resulting in circular polarised radiation. In **Figure 25(b)**, a 2 × 2 sequentially rotated array is shown. In this arrangement, there exists two set of orthogonal patches (patches 1 and 3 and patches 2 and 4) that generate the two orthogonal modes with 90° phase difference leading to gain enhancement compared to the antenna shown in **Figure 25(a)**.

In order to increase the gain more, the number of elements should be increased. In Ref. [23], it is shown that for this purpose, if N elements are to be used, then they should be arranged in a circular ring and the rotation angle of each element relative to its neighbouring element should be 360/N degree. Furthermore, the phase shifting between the neighbouring elements should be 360/N as well.

8. Other techniques

In this section, some other techniques that are reported in the literature are introduced and briefly described. However, as we know there exist numerous works on bandwidth enhancement of microstrip patch antennas, with respect to all researchers who have worked on this issue, we have selected a few number of works and presented here.

8.1. Log-periodic array of patches

In [24–26], the idea of log-periodic antennas for providing very wide bandwidths has been applied to microstrip patches. In this technique, the microstrip patches are arranged in a log-periodic formation. The patches are series fed by a microstrip line either directly or indirectly. In **Figure 26(a** and **b)**, both direct- and indirect-fed configurations are illustrated. In the configuration shown in **Figure 26(a)**, each patch is coupled to the feed line located at the lower layer; however in **Figure 26(b)**, each patch is directly connected to feed line via a quarter-wave transformer for easier impedance matching (i.e. the length 'd' is chosen $\lambda/4$ where λ is the wavelength corresponding to the resonance frequency of the patch). In **Figure 26(c)**, the VSWR plot of a sample antenna designed in Ref. [26] based on the configuration shown in **Figure 26(b)**, is shown. It can be observed that very wide bandwidth (about 100%) is obtained by making use of this configuration.

The radiation pattern of antennas shown in **Figure 26** is broadside, unlike the conventional log-periodic dipole array which radiates at end-fire direction. Although by using log-periodic configuration, very wide bandwidths can be achieved, but the main beam direction scans vs. frequency makes it impossible to provide constant beam direction through the whole bandwidth.

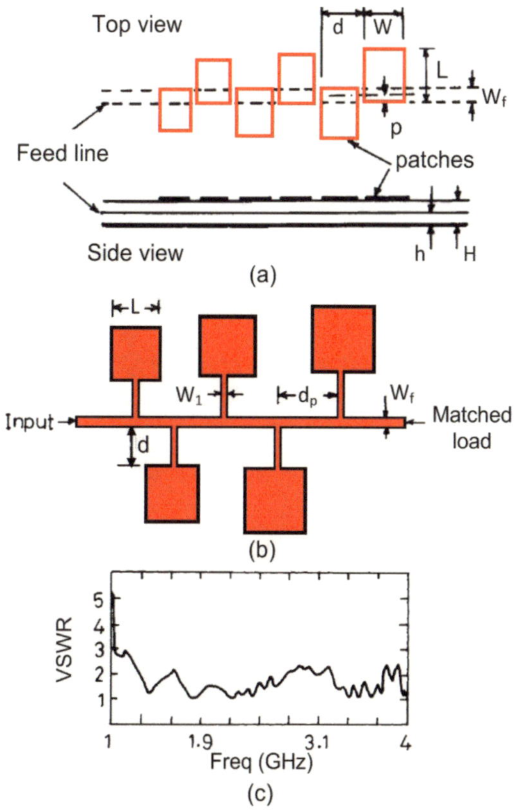

Figure 26. Microstrip patches arranged in log-periodic array formation. (a). Indirect feeding, (b). direct feeding and (c). VSWR plot of a sample antenna designed in Ref. [26] based on the configuration shown in **Figure 29(b)**.

8.2. E-shaped patch

In Ref. [27], an E-shaped patch backed by a SIW[1] cavity is proposed by which 10.9% bandwidth could be obtained. The antenna structure is illustrated in **Figure 27**. As shown, the radiating element is an E-shaped patch backed by SIW cavity and fed by a strip line which is loaded by some stubs. Direct contact between feed line and radiating patch is provided by a metalized via at the centre performing like a probe.

In Ref. [28], it is declared that for a conventional E-shaped patch (i.e. a single layered E-shaped patch antenna fed by a simple probe) in order to provide impedance matching for the lower resonating mode, we need to choose the substrate thickness more than 0.07λ, since the inductance contributed by the probe is very low and the capacitance introduced by the patch and ground becomes very large [27]. In this design, some stubs are introduced at the end of strip line in order to provide more inductance and make it possible to impedance match the lower resonating mode even with thin substrates of <0.07λ thicknesses.

[1]Substrate integrated waveguide.

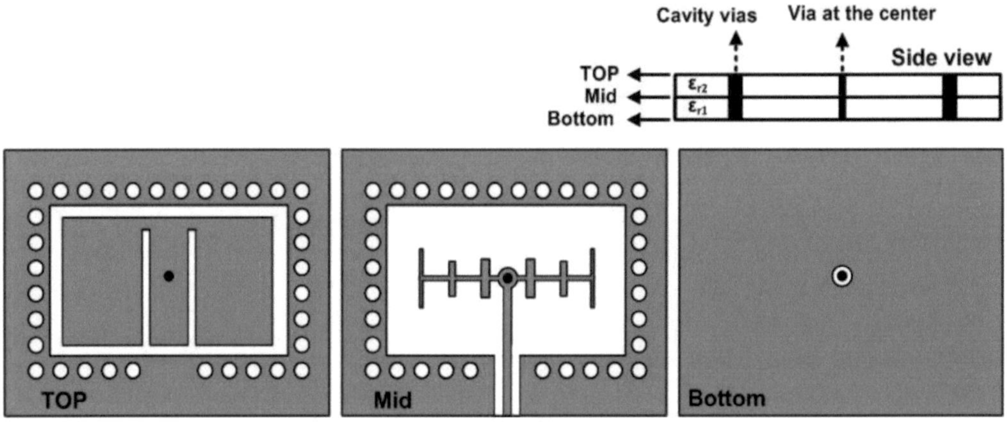

Figure 27. The E-shaped patch antenna configuration presented in Ref. [27].

The input impedance and current distributions corresponding to the two resonating modes of radiating patch for the two cases, with and without stubs, are depicted in **Figure 28**. It can be clearly observed that if the stubs are not used, the lower resonating frequency cannot be created with the thin substrate used in this design. However, by applying the stubs and fine tuning of their dimensions, two resonating modes that are successfully impedance matched make it possible to increase the impedance bandwidth. In **Figure 29**, the reflection coefficient, S11, of the antenna designed in Ref. [27] is shown and the impedance bandwidth (S11 < −10 dB) of 10.9% (9.45–10.54 GHz) can be observed.

8.3. L-shaped feeding

In [29, 30], an L-shaped probe is used to feed a microstrip patch and demonstrated that by this way, the impedance bandwidth can be increased. This method can be interpolated as a

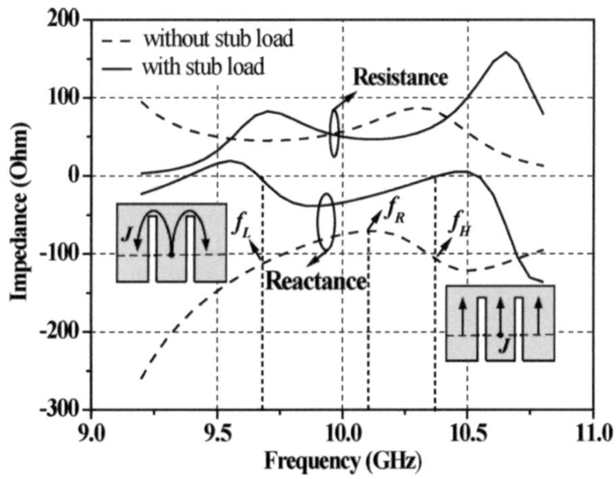

Figure 28. The input impedance and current distributions corresponding to the two resonating modes of radiating patch for the two cases, with and without stubs, for the antenna shown in **Figure 27** [27].

http://dx.doi.org/10.5772/intechopen.70173

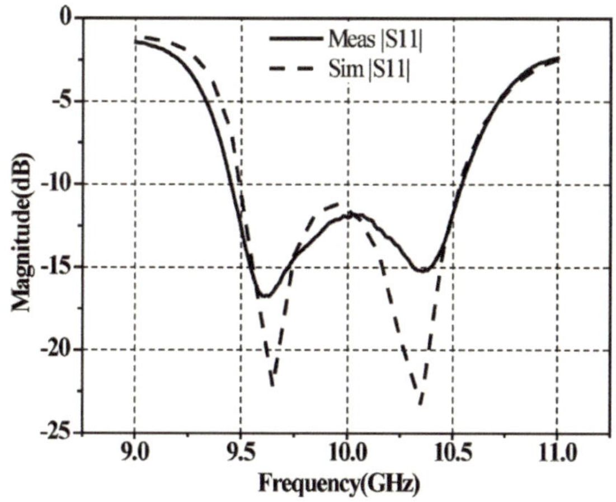

Figure 29. Reflection coefficient, S11, of the sample antenna designed in Ref. [27] based on the configuration shown in **Figure 27** [27].

combination of proximity coupled feeding and stacking technique. In **Figure 30**, a rectangular patch antenna fed by an L-shaped probe is illustrated. As shown, the L-shape probe is realised by connecting a metalized via to the feed line. In this figure, the L-shaped probe is fed by a strip line in order to supress parasitic radiation due to the feed structure. The simulated S11 of a sample designed in Ref. [30] is shown in **Figure 31**. It can be seen that 50.4% (46–77 GHz) impedance bandwidth is achieved.

8.4. Microstrip monopole slotted antenna

In Ref. [31], a microstrip monopole slotted antenna is introduced and its performance using three different shapes (straight, L and inverted-T) for the slot is investigated and it is shown

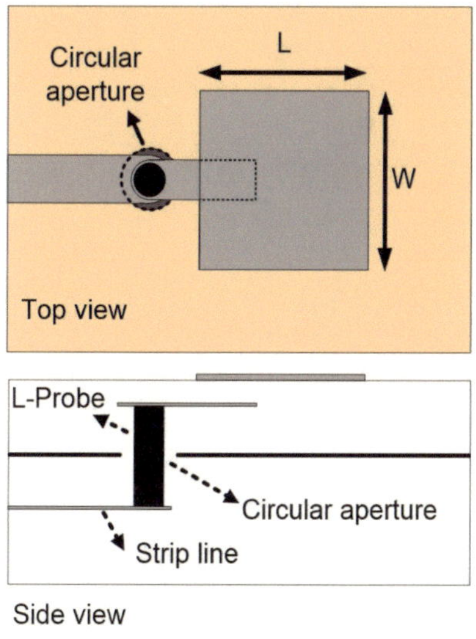

Figure 30. Rectangular patch antenna fed by an L-shaped probe [30].

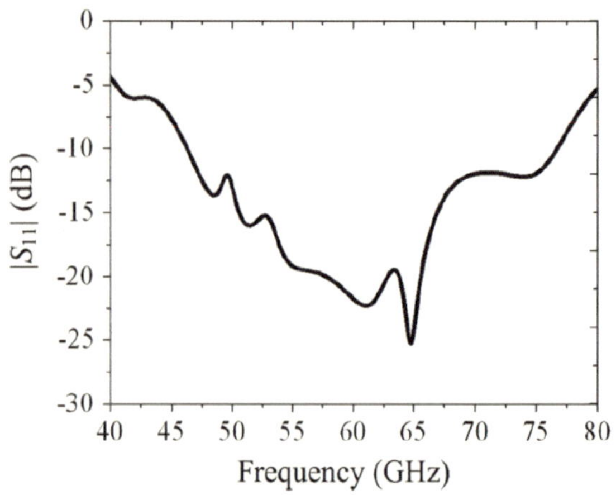

Figure 31. The simulated S11 of a sample antenna designed in Ref. [30] based on the configuration shown in **Figure 30**.

that by making use of this single-layer configuration, a wide impedance bandwidth of more than 80% can be achieved. It should be said that this antenna has omnidirectional radiation pattern and consequently lower gain compared to the antennas presented in previous sections.

In **Figure 32**, the antenna configuration for three different slot shapes is shown. As depicted, it is a single-layer structure in which the slot is electromagnetically coupled to the feed line. The presence of slot introduces additional resonances that can be well-coupled to the patch resonance leading to antenna bandwidth enhancement.

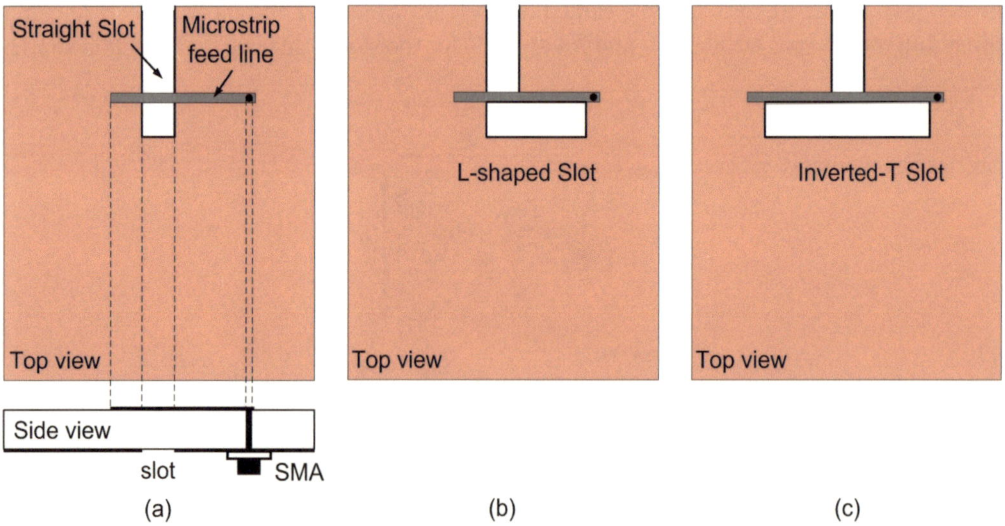

Figure 32. Microstrip monopole slotted antenna with three different slot shapes: (a). straight slot, (b). L-shaped slot and (c). inverted-T slot [31].

In Ref. [31], three samples of antennas shown in **Figure 32** are designed. **Figure 33** shows the reflection coefficients, S11, of the designed antennas. It can be clearly observed that for the case a single straight slot is used (**Figure 32(a)**), one additional resonance is created resulting in 57% bandwidth (2.5–4.5 GHz); however, for the cases where L-shaped (**Figure 32(b)**) or inverted-T slot (**Figure 32(c)**) is used, three additional resonances are created resulting in wider impedance bandwidth: 82% (2.42–5.78 GHz) for the former and 80% (2.74–6.4 GHz) for the latter. The reason is that for the cases of L-shaped and inverted-T slots, the two orthogonal arms of slot perform as two separate resonators properly coupled to each other and the patch making four well-coupled resonances which leads to significant enhancement of the bandwidth.

8.5. Defected ground/patch

Defecting the patch/ground has been used as a technique to improve the radiation characteristics of microstrip patch antennas. Mutual coupling suppression in arrays [32–34], improving the efficiency [35], reducing the antenna size [36] and lowering the cross-polar level [37–39] and Bandwidth enhancement [40–42] are some examples of this. Here, our focus is on the latter one, i.e. the bandwidth enhancement.

In Ref. [40], a double U-shaped slot is introduced to the ground plane of a microstrip monopole antenna in order to enhance its impedance bandwidth. With the resultant structure, the impedance bandwidth of 114% was obtained. The proposed structure and reflection coefficient S_{11} of a designed sample in Ref. [40] are illustrated in **Figure 34**.

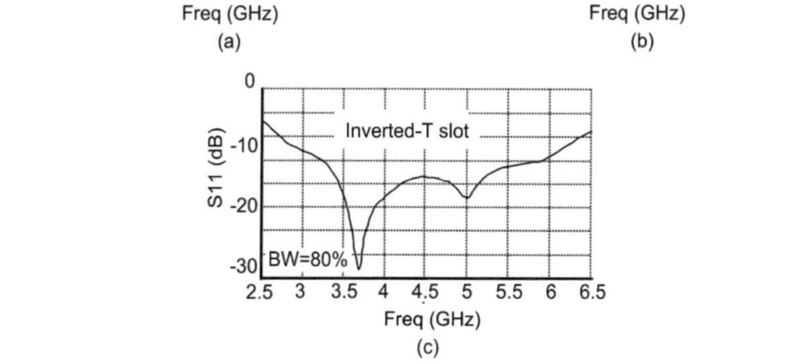

Figure 33. The reflection coefficients, S11, of the sample antennas designed in Ref. [31] based on the configurations shown in **Figure 32**. (a). Straight slot, (b). L-shaped slot and (c). inverted-T slot [31].

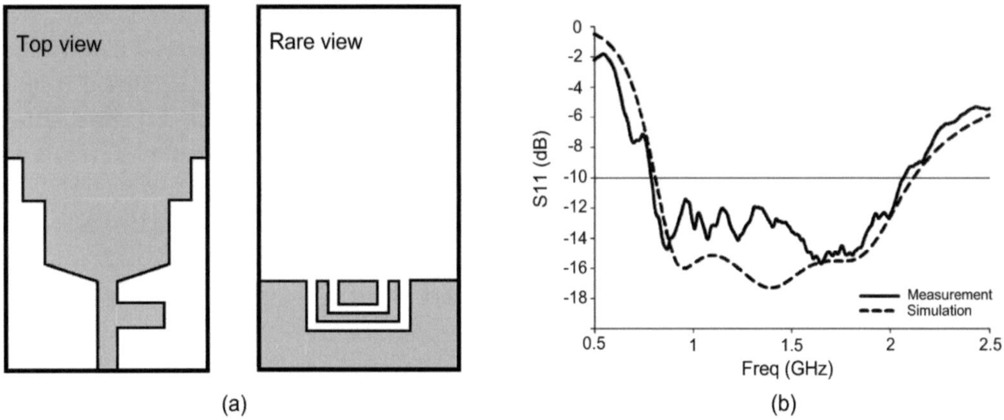

Figure 34. The proposed defected ground microstrip monopole antenna presented in Ref. [40]. (a). Topology of the structure. (b). Reflection coefficient S_{11} of a designed sample in Ref. [40].

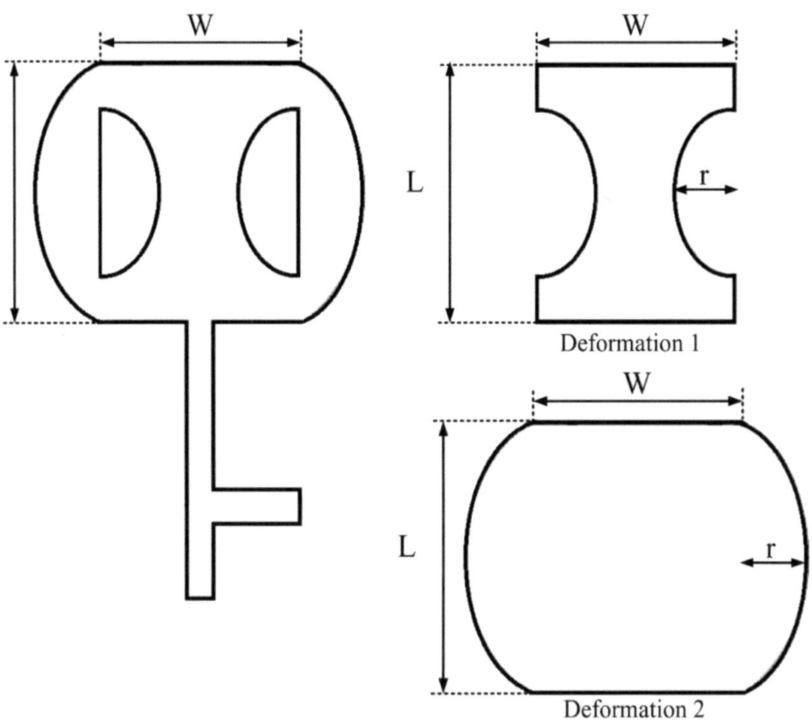

Figure 35. The proposed D-shaped defected microstrip patch presented in Ref. [41]. Complete structure and two deformations used in it [41].

In Ref. [41], a D-shaped defected patch is used for bandwidth enhancement. The proposed structure and the two deformations used in it are shown in **Figure 35**. By deformation 1, the resonant frequency of TM_{10} mode decreases since the effective length of the patch increases for this resonating mode. However, the deformation 2 performs inversely, i.e. it decreases the effective length of the patch for TM_{10} mode increasing its resonant frequency. As a result, combination of these two deformations makes a D-shaped defected patch which can produce two resonant frequencies by the two effective lengths of inside and outside of D-shaped defection. The open circuit stub connected to the feed line is used for impedance matching.

In **Figure 36**, a fabricated sample of proposed structure and its reflection coefficient S_{11} are depicted. The reflection coefficient of the proposed structure is also compared with that of a conventional patch and 5% bandwidth enhancement can be observed in the measurement result.

In Ref. [42], a dumbbell-shaped defected patch is used to both increase the bandwidth and enhance polarisation purity. The proposed structure is shown in **Figure 37**. As shown, a pair of thin slots (with the dimensions $l_1 \times S_{w1}$) is etched near the non-radiating edges and four wider slots (with the dimensions $l_2 \times d_w$) are positioned near the patch corners. By this way, the fields corresponding to non-radiating edges and corners which mainly contribute to cross-polar radiation are perturbed leading to more pure radiations. In Ref. [42], it is shown that each $l_1 \times S_{w1}$ slot introduces a reactance parallel to the input impedance of the patch in such a way that the variation of patch input reactance vs. frequency (as shown in **Figure 38**) becomes smaller compared to the conventional patch. This fact may point to the broader bandwidth of the proposed structure relative to the conventional patch.

In **Figure 39**, the reflection coefficient of a sample antenna designed in Ref. [42] is illustrated and compared with that of a conventional patch. Results show that the proposed antenna produces broader bandwidth. The measurement result shows 16% bandwidth for the designed sample.

8.6. Recent works

In this part, we have tried to provide a brief review of the most important works that has been done in recent years to give the reader a prospective of the latest researches on bandwidth enhancement of microstrip patch antennas.

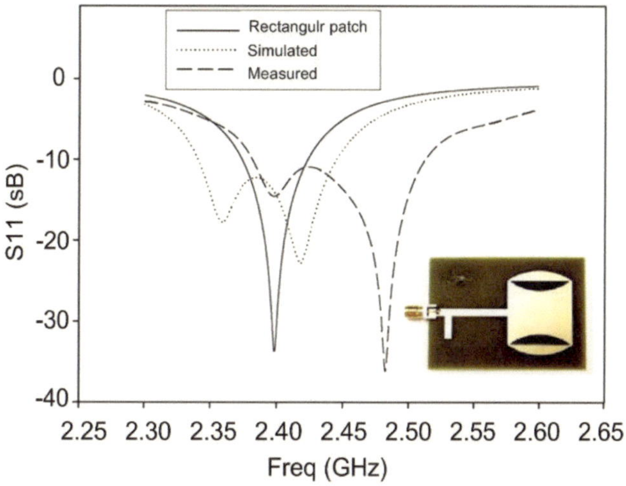

Figure 36. Fabricated sample of D-shaped defected microstrip patch presented in Ref. [41] and comparison of its reflection coefficient with that of a conventional patch.

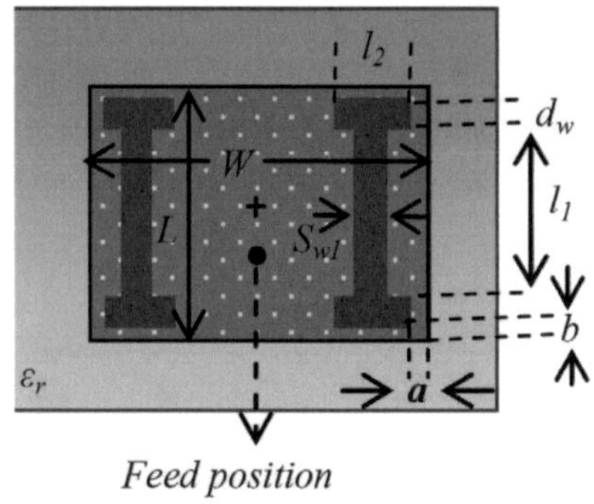

Figure 37. The proposed dumbbell-shaped defected patch presented in Ref. [42].

The latest works focused on the enhancement of the bandwidth without increasing the size and height of antenna [43–46]. In [43–45], differential feeding is used to suppress the excitation of undesired modes. Then by joining the desired modes (or by introducing extra modes between the desired modes), the bandwidth for broadside radiation is enhanced.

In Ref. [43], by differential feeding for a rectangular patch, the unwanted modes TM_{20}, TM_{12} and TM_{22} are supressed and the favourite broadside modes TM_{10} and TM_{30} are excited. Then by making use of two stubs connected to the radiating edges of the patch and a stepped ground plane, as shown in **Figure 40(a)**, it became possible to enclose the desired modes leading to bandwidth enhancement. In **Figure 40(b)**, the reflection coefficient of a sample designed in Ref. [43] is shown and about 10% bandwidth is observed.

In Ref. [44], as shown in **Figure 41(a)**, a differentially fed equilateral triangular patch is loaded by shorting posts. In this structure, the desired modes TM_{10} and TM_{11} are well excited using differential feeding. Loading the patch with shorting pins generates additional mode, i.e. zero mode, between the two favourite modes which makes it possible to enhance the antenna

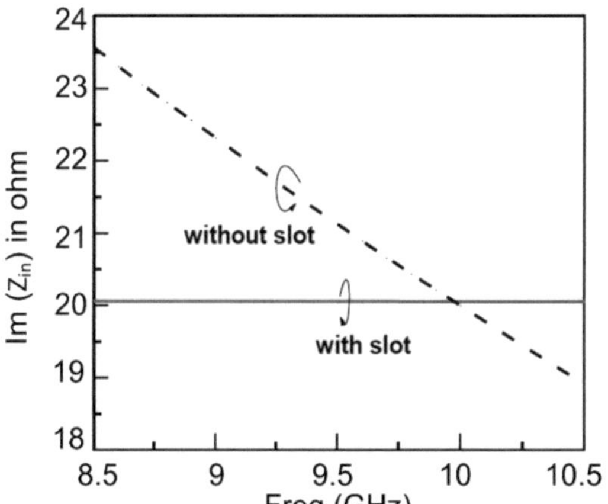

Figure 38. Variation input reactance of proposed patch in Ref. [42] vs. frequency and its comparison with that of a conventional patch.

http://dx.doi.org/10.5772/intechopen.70173

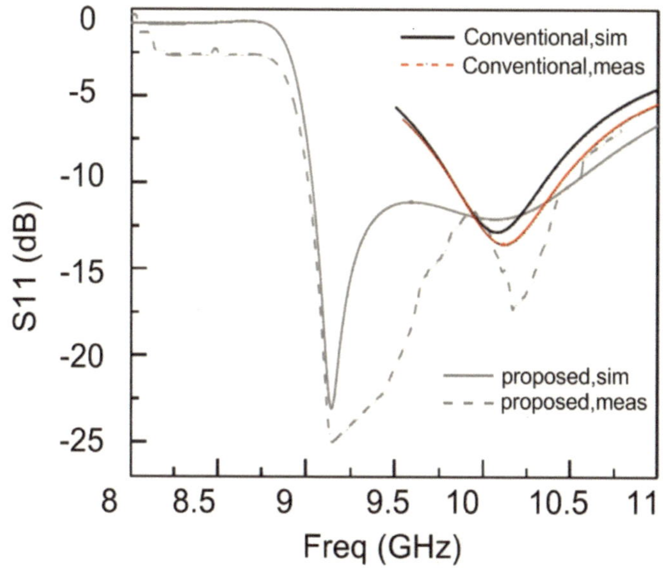

Figure 39. Reflection coefficient of a sample antenna designed in Ref. [42] and its comparison with that of a conventional patch.

bandwidth by making a proper coupling between the three modes. It should be noted that the shorting pins are located so that they do not perturb the field distribution of TM_{10} and TM_{11} modes and consequently do not affect them. In **Figure 41(b)**, the reflection coefficient of a sample antenna designed in Ref. [44] is shown and it is observed that more than 50% bandwidth could be obtained.

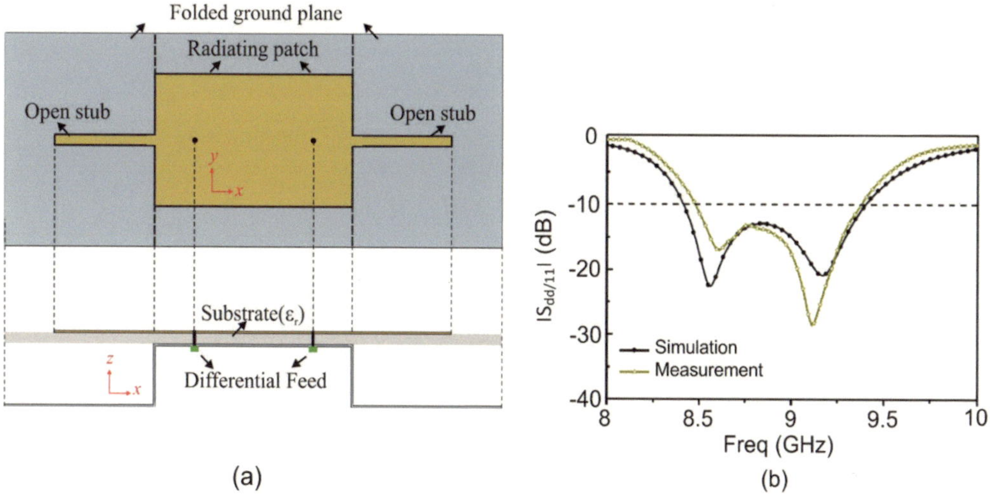

Figure 40. The differentially fed rectangular patch antenna presented in Ref. [43]. (a). Topology of the structure. (b). Reflection coefficient of a sample designed in Ref. [43].

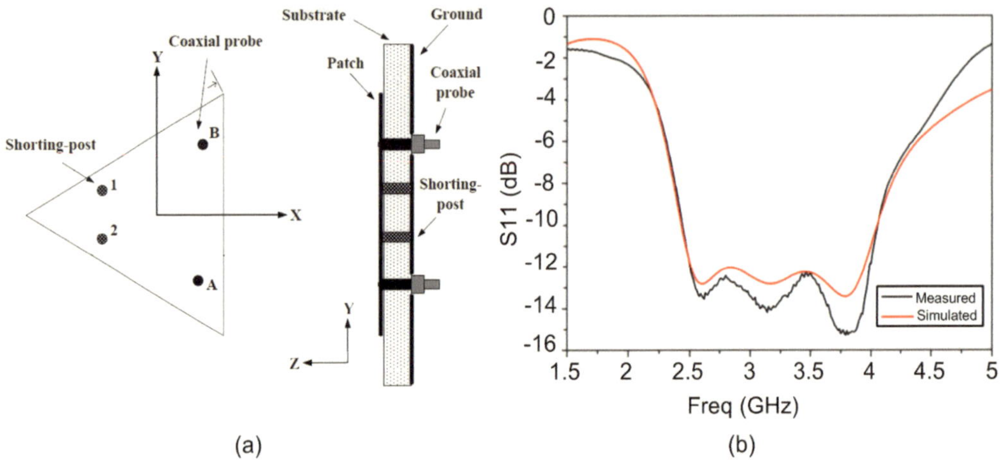

Figure 41. The differentially fed equilateral triangular patch antenna loaded by shorting posts which is presented in Ref. [44]. (a). Topology of the structure. (b). Reflection coefficient of a sample antenna designed in Ref. [44].

In Ref. [45], in the same way as presented in Ref. [43], a rectangular patch is differentially fed so that the spurious modes between TM_{10} and TM_{30} modes cannot be excited. But here these two modes are coupled to each other by another way. The proposed structure in Ref. [45] is shown in **Figure 42(a)**. In this structure, the shorting pins increase the resonant frequency of TM_{10} mode without any effect on TM_{30} mode; however, the pair of long slots shifts down the resonant frequency of TM_{30} mode with slight effect on TM_{10} mode. The short centred slot is also used for better impedance matching. By this way, the desired modes TM_{10} and TM_{30} could be well-coupled leading to enhancement of the antenna bandwidth. In **Figure 42(b)**, the reflection coefficient of a sample antenna designed in Ref. [45] is shown and about 13% bandwidth can be observed.

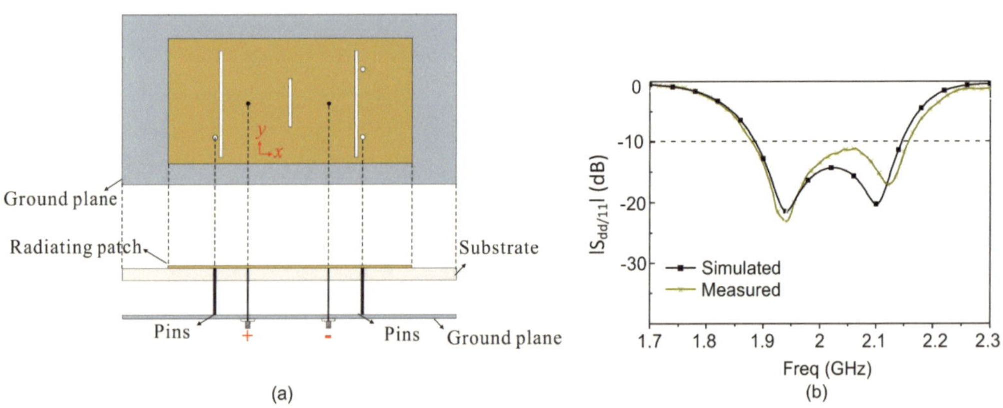

Figure 42. The differentially fed rectangular patch antenna presented in Ref. [45]. (a). Topology of the structure. (b). Reflection coefficient of a sample designed in Ref. [45].

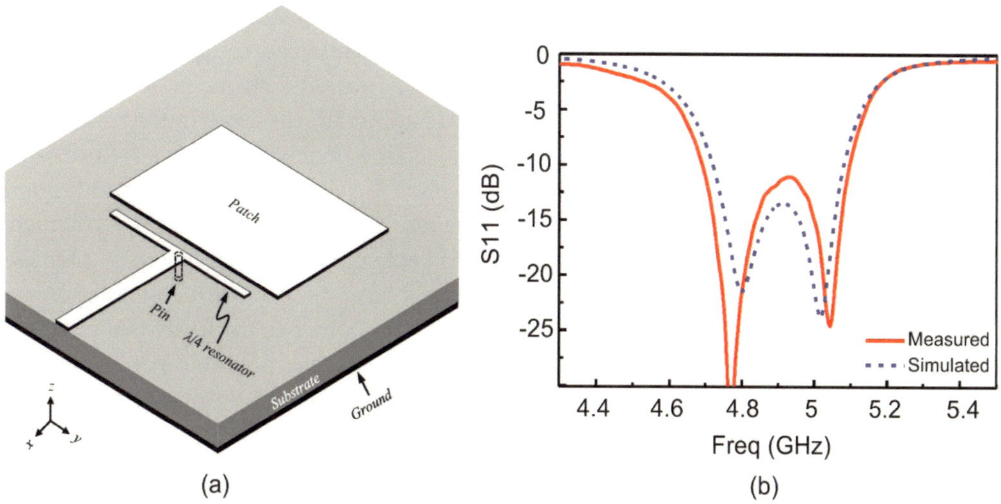

Figure 43. The capacitively fed rectangular patch antenna presented in Ref. [46]. (a). Topology of the structure. (b). Reflection coefficient of a sample designed in Ref. [46].

In Ref. [46], as shown in **Figure 43(a)**, the microstrip patch is not fed directly. In this structure, the feed line excites a pair of non-radiating quarter-wavelength microstrip line resonators which are capacitively coupled to the patch. As a result, dual-resonance structure is provided in which the two resonances correspond to the patch and quarter-wavelength microstrip line resonators, respectively. By tuning the gap between the resonators and the patch, a proper coupling between them and consequently wider impedance bandwidth can be obtained. In addition, the shorting pin introduced in the central plane suppresses the excitation of all even-order modes in quarter-wavelength resonators and consequently in the patch. As a result, the TM_{20} mode which is the most harmful mode that can generate cross-polar radiation is naturally supressed. In **Figure 43(b)**, the reflection coefficient of a sample antenna designed in Ref. [46] is shown by which the impedance bandwidth of about 8.5% bandwidth could be obtained.

9. Summary

In this chapter, a variety of procedures proposed in the literature to increase the impedance bandwidth of microstrip patch antennas are presented and discussed. In Section 2, bandwidth enhancement by proper choosing of substrate is described and declared that up to 10% bandwidth can be obtained using this method taking this fact into consideration that other characteristics of antenna, say efficiency and backward radiation, may be affected. In Section 3, it is shown that by applying indirect feeding mechanism and coupling the electromagnetic fields to the patch, bandwidth can be enhanced. Two methods including proximity coupled and aperture-coupled patches were introduced by which up to 10% and

30% bandwidth can be achieved, respectively. However, for proximity coupled patches, if external matching structure is used, the achievable bandwidth can be doubled. In Section 4, bandwidth enhancement making use of horizontally parasitic elements at the antenna aperture is described and stated that namely 20% bandwidth can be achieved using this single-layer technique at the expense antenna size enlargement which makes it unsuitable for use in array configurations. Sections 5 and 6 explore the stacked patches, a multilayered solution using vertically parasitic elements that can result in very wide bandwidths (namely 50–70%). By this method, the antenna planar size is kept small making it interesting for use in arrays; however, due to its multilayered topology, it has problems with cost and fabrication complexity compared to single-layer solutions. In Sections 2–6, the focus was on linear polarised antennas. Therefore, the solutions for wideband circular polarised patch antennas are introduced and discussed in Section 7. The examples presented in this section reveals that namely up to 20% AR bandwidth can be achieved using the techniques introduced in Section 7. Finally, in Section 8, other techniques proposed in the literature including log-periodic array of patches, E-shaped patch, L-shaped feeding, microstrip monopole slotted antenna, defected ground/patch technique and the latest works in recent years are introduced and investigated. The bandwidth obtained by some designed samples using these structures were 100%, 11%, 50%, 82%, 114% and 50%, respectively.

Author details

Seyed Ali Razavi Parizi

Address all correspondence to: alirazaviparizi@gmail.com

Department of Electrical and Computer Engineering, Graduate University of Advanced Technology, Kerman, Iran

References

[1] Waterhouse RB. Microstrip Patch Antennas: A Designer's Guide. Springer Science & Business Media; New York; 2003

[2] Kumar G, Ray KP. Broadband Microstrip Antennas. Artech House; Boston; 2003

[3] Zurcher JF. The SSFIP: A global concept for high performance broadband planar antennas. Electronics Letters. 1988;**24**:1433-1435

[4] Rathi V, Kumar G, Ray KP. Improved coupling for aperture coupled microstrip antennas. IEEE Transactions on Antennas and Propagation. 1996;**44**(8):1196-1198

[5] Kumar G, Gupta KC. Broadband microstrip antennas using additional resonators gap-coupled to the radiating edges. IEEE Transactions on Antennas and Propagation. 1984;**32**: 1375-1379

[6] Kumar G, Gupta KC. Nonradiating edges and four edges gap-coupled with multiple resonator, broad band microstrip antennas. IEEE Transactions on Antennas and Propagation. 1985;**33**:173-178

[7] Chew WC. A broadband annular ring microstrip antenna. IEEE Transactions on Antennas and Propagation. 1982;**30**:918-922

[8] Targonski SD, Waterhouse RB, Pozar DM. Design of wideband aperture stacked patch microstrip antenna. IEEE Transactions on Antennas and Propagation. 1998;**46**(9):1245-1251

[9] Reddy KTV, Kumar G. Dual feed gap coupled square microstrip antennas for broadband circularly polarization. Microwave and Optical Technology Letters. 2000;**26**(6):399-402

[10] Reddy KTV, Kumar G. Gap Coupled Broadband Circularly Polarized Square Microstrip Antennas. In: International Conference on Computers, Communication and Devices, ICCCD; 2000. p. 365-368

[11] Reddy KTV, Kumar G. Stacked square microstrip antennas for wideband circular polarization. In: National Conf. on Communications, NCC. Kanpur: Indian Institute of Technology; 2001. p. 125-128

[12] Huang CY, Wu JY, Wong KL. Cross-slot-coupled microstrip antenna and dielectric resonator antenna for circular polarization. IEEE Transactions on Antennas and Propagation. 1999;**47**(4):605-609

[13] Vlasits T. Performance of a cross aperture coupled single feed circularly polarized patch antenna. Electronic Letters. 1996;**32**(7):612-613

[14] Targonski SD, Pozar DM. Design of wideband circularly polarized aperture coupled microstrip antennas. IEEE Transactions on Antennas and Propagation. 1993;**41**(2):214-220

[15] Pozar DM, Duffy SM. A dual-band circularly polarized aperture-coupled stacked microstrip antenna for global positioning satellite. IEEE Transactions on Antennas and Propagation. 1997;**45**(11):1618-1625

[16] Aksun MI, Chuang SL, Lo YT. On slot-coupled microstrip antennas and their applications for circular polarization operation. IEEE Transactions on Antennas and Propagation. 1990;**38**:1224-1230

[17] Huang CY, Wu CY, Wong KL. Slot-coupled microstrip antenna for broadband circular polarization. Electronic Letters. 1998;**34**(9):835-836

[18] James JR, Hall PS. Handbook of Microstrip Antennas. Vol. 1. London: Peter Peregrinus; 1989

[19] Pozar DM, Schaubert DH. Microstrip Antennas: The Analysis and Design of Microstrip Antennas and Arrays. New York: IEEE Press; 1995

[20] Teshirogi TM, Chujo W. Wideband Circularly Polarized Antenna with Sequential Rotations and Phase Shift of Elements. In: ISAP; 1985. p. 117-120

[21] Lo WK, Chan CH, Luk KM. Circularly polarized microstrip antenna Array using proximity coupled feed. Electronic Letters. 1998;**34**(23):2190-2191

[22] Huang J. A technique for an array to generate circular polarization with linear polarized elements. IEEE Transactions on Antennas and Propagation. 1986;**34**(9):1113-1123

[23] Iwasaki H, Nakajima T, Suzuki Y. Gain improvement of circularly polarized array antenna using linearly polarized elements. IEEE Transactions on Antennas and Propagation. 1995; **43**(6):604-608

[24] Hall PS. Multi-octave bandwidth log-periodic microstrip antenna array. IEE Proceedings—Microwaves, Antennas and Propagation. 1986;**133**(2):127-136

[25] Pues H. Wideband quasi log-periodic microstrip antennas. IEE Proceedings—Microwaves, Antennas and Propagation. 1981;**128**:159-163

[26] Kakkar R, Kumar G. Stagger tuned microstrip log-periodic antenna. IEEE Antennas and Propagation Society International Symposium. Digest. 1996:1262-1265

[27] Yang W, Zhou J. Wideband low-profile substrate integrated waveguide cavity-backed E-shaped patch antenna. IEEE Antennas and Wireless Propagation Letters. 2013;**12**:143-146

[28] Awida MH, Suleiman SH, Fathy AE. Substrate-integrated cavity-backed patch arrays: A low-cost approach for bandwidth enhancement. IEEE Transactions on Antennas and Propagation. 2011;**59**(4):1155-1163

[29] Kuo l-S, Wong K-L. A dual-frequency L-shaped patch antenna. Microwave and Optical Technology Letters. 2000;**27**:177-179

[30] Wang L, Guo YX, Sheng WX. Wideband high-gain 60-GHz LTCC L-probe patch antenna array with a soft surface. IEEE Transactions on Antennas and Propagation. 2013;**61**(4): 1802-1809

[31] Latif SI, Shafai L, Sharma SK. Bandwidth enhancement and size reduction of microstrip slot antennas. IEEE Transactions on Antennas and Propagation. 2005;**53**(3):994-1003

[32] Guha D, Biswas S, Biswas M, Siddiqui JY, Antar YMM. Concentric ring shaped defected ground structures for microstrip circuits and antennas. IEEE Antennas and Wireless Propagation Letters. 2006;**5**:402-405

[33] Salehi M, Tavakoli A. A novel low mutual coupling microstrip antenna array design using defected ground structure. International Journal of Electronics and Communications. 2006;**60**:718-723

[34] Guha D, Biswas S, Joseph T, Sebastian MT. Defected ground structure to reduce mutual coupling between cylindrical dielectric resonator antennas. Electronic Letters. 2008;**44**(14): 836-837

[35] Arya AK, Kartikeyan MV, Patnaik A. Efficiency enhancement of microstrip patch antenna with defected ground structure. In: International Conference on Recent Advances in Microwave Theory and Applications, Microwave; 2008

[36] Zheng SY, Yeung SH, Chan WS, Man KF, Leung SH. Size-reduced rectangular patch hybrid coupler using patterned ground plane. IEEE Transactions on Microwave Theory and Techniques. 2009;**57**(1):180-188

[37] Kumar C, Guha D. Nature of cross-polarized radiation from probe fed circular microstrip antenna and their suppression using different geometries of DGS. IEEE Transactions on Antennas and Propagation. 2012;**60**:92-101

[38] Kumar C, Guha D. DGS integrated rectangular microstrip patch for improved polarization purity with wide impedance bandwidth. IET Microwave Antennas Propagation. 2014;**8**(8):589-596

[39] Ghosh A, Ghosh D, Chattopadhyay S, Singh LLK. Rectangular microstrip antenna on slot type defected ground for reduced cross polarized radiation. IEEE Antennas and Wireless Propagation Letters. 2015;**14**:321-324

[40] Chiang KH, Tam KW. Microstrip monopole antenna with enhanced bandwidth using defected ground structure. IEEE Antennas and Wireless Propagation Letters. 2008;**7**: 532-535

[41] You W, Guo H, Cai W, Liu X. A D-shaped defected patch antenna with enhanced bandwidth. IEEE International Symposium on Microwave, Antenna, Propagation and EMC Technologies for Wireless Communications. 2009:684-686

[42] Chakraborty S, Ghosh A, Chattopadhyay S, Singh LK. Improved cross polarized radiation and wide impedance bandwidth from rectangular microstrip antenna with dumbbell shaped defected patch surface. IEEE Antennas and Wireless Propagation Letters. 2016;**15**:84-88

[43] Liu NW, Zhu L, Choi WW, Zhang JD. A novel differential-fed patch antenna on stepped-impedance resonator with enhanced bandwidth under dual-resonance. IEEE Transactions on Antennas and Propagation. 2016;**64**(11):4618-4625

[44] Wang J, Liu Q, Zhu L. Bandwidth enhancement of a differential-fed equilateral triangular patch antenna via loading of shorting posts. IEEE Transactions on Antennas and Propagation. 2017;**65**(1):36-43

[45] Liu NW, Zhu L, Choi WW. A differential-fed microstrip patch antenna with bandwidth enhancement under operation of TM10 and TM30 modes. IEEE Transactions on Antennas and Propagation. 2017;**65**(4):1607-1614

[46] Zhang JD, Zhu L, Wu QS, Liu NW, Wu W. A compact microstrip-fed patch antenna with enhanced bandwidth and harmonic suppression. IEEE Transactions on Antennas and Propagation. 2016;**64**(12):5030-5037

Chapter 2

Miniature Planar Antenna Design for Ultra-Wideband Systems

Mohammad Alibakhshikenari,
Mohammad Naser-Moghadasi,
Ramazan Ali Sadeghzadeh, Bal Singh Virdee and
Ernesto Limiti

Additional information is available at the end of the chapter

http://dx.doi.org/10.5772/intechopen.68612

Abstract

Demand for antennas that are compact and operate over an ultra-wideband (UWB) frequency range is growing rapidly as UWB systems offer high resolution imaging capability and high data rate transmission in the order of Gb/s that is required by the next generation of wireless communication systems. Hence, over the recent years the research and development of UWB antennas has been widely reported in literature. The main performance requirements sought from such antennas include: (1) low VSWR of <2; (2) operation over 7.6 GHz from 3 to 10.6 GHz; and (3) good overall radiation characteristics. Significant size reduction and low manufacturing cost are also important criteria in order to realize a cost-effective and miniature system. Other desirable requirements include compatibility and ease of integration with RF electronics.

Keywords: ultra-wideband antenna, metamaterial, composite right/left-handed transmission line, antenna miniaturization, slit antennas

1. Introduction

Ever since the Federal Communications Commission (FCC) released a bandwidth of 7.5 GHz (from 3.1 to 10.6 GHz) for ultra-wideband (UWB) wireless communications, UWB technology has rapidly developed for high data rate wireless communications [1–5]. As is the case in conventional narrowband wireless communication systems, an antenna plays a very crucial role in UWB systems. However, there are greater challenges in designing a UWB antenna than a narrow band one. A suitable UWB antenna should be capable of operating over an ultra-wide bandwidth as defined by the FCC. At the same time, satisfactory radiation properties over the UWB frequency range are also necessary, including nondispersive nature in order to minimize distortion in the transmitted signal. The antenna needs to have a low voltage standing wave ratio (VSWR) (<2) over 3.1–10.6 GHz band and omnidirectional radiation characteristics [6]. Nowadays, in most applications significant size reduction of antennas is paramount in order to achieve minimization of communication systems. Other desirable features include being a planar structure that is cost-effective to fabricate in large volumes.

Design of a UWB antennas is challenging for systems operating at the lower part of the microwave spectrum. This is due to the fact that the wavelength is very large at these frequencies and the resulting physically large antennas are not desirable due to space limitations of modern systems. This problem is particularly severe in systems that operate at HF, VHF, and UHF bands. This is because in such applications having antennas with low visual signatures is of paramount importance. The current antennas of choice for these applications tend to be monopole whip antennas. These antennas, however, suffer from two major drawbacks. First, the relatively large heights of a conventional monopole whip antenna, when mounted on a vehicle, such an antenna significantly protrudes from the top surface of the vehicle drastically increasing the visual signature of the vehicle. The second issue with monopole whip antennas is their narrow bandwidths, which limits the types of waveforms that they can receive or transmit. Therefore, development of compact, low-profile, and ultra-wideband antennas is of particular interest in many communications systems that operate at HF, VHF, and UHF frequencies. To increase the bandwidth of monopole-type radiators, a number of different techniques have been examined. A variety of printed monopole antennas that provide UWB operation in a planar form factor are examined in the 3.1–10.6 GHz band [7–9]. However, at its lowest frequency of operation, a printed monopole tends to have relatively large dimensions.

Numerous techniques have been investigated and reported in recent years to reduce the size of microstrip antennas. These techniques are mainly based on loading a patch antenna with reactive components realized with suitably designed slots, shorting posts, and lumped elements. The effectiveness of these techniques is, however, limited in the reduction of the footprint of planar antennas which is required by modern wireless systems [10–12]. The size of patch antennas can also be reduced by fabricating the antenna on dielectric substrates with a high permittivity. This well-established technique, however, results in increased surface wave excitation over the antenna that degrades its impedance bandwidth, radiation efficiency, and its radiation characteristics [13].

Techniques mentioned above fail to meet the challenges for miniaturization of antennas. In this chapter, the exploitation of artificially engineered materials that are based on metamaterial transmission lines is shown to provide the solution for miniaturization of planar antennas [14–22]. In fact metamaterials, which are also accurately referred to as composite right-/left-handed (CRLH) transmission lines, are a novel paradigm in electromagnetic engineering as such materials exhibit electrical characteristics, that is, negative permittivity and permeability, not possible with naturally occurring materials [11]. These properties are exploited in this chapter to design planar antennas with a small footprint using standard manufacturing photolithography techniques [11–18]. In particular, in this chapter, a unique CRLH-TL structure is employed in the design of an antenna using conventional microwave integrated

http://dx.doi.org/10.5772/intechopen.68612

circuit (MIC) manufacturing techniques. The unit cells of the CRLH-TL structure was realized by engraving L- and T-shaped slits on a rectangular conductor that is grounded through a high impedance microstrip stub that are spiraled. In the antenna the L- and T-shaped slits exhibit capacitive property, and the stub acts like an inductor. The antenna structure, which was created using these unit cells, was modeled and optimized for UWB performance while maintaining its radiation characteristics in terms of gain and radiation efficiency.

2. Composite right-/left-handed antenna design

A left-handed transmission line structure can be created from an arrangement consisting of a series capacitor and shunt inductor. This *LC* circuit configuration can be achieved for antenna applications by simply etching a dielectric slot in a metallic-radiating patch, where the patch is grounded using a high impedance microstrip line. The slot created in the patch acts like a series capacitance C_L and the high impedance microstrip line acts like a shunt inductance L_L. The inductive element in the structure can be coiled to reduce the overall footprint of the antenna structure [23, 24]. This approach is used here to implement a compact antenna by cascading together a number of CRLH transmission line unit cells comprising left-handed *LC* unit cells.

The equivalent circuit model for a lossless CRLH transmission line unit cell is shown in **Figure 1**. It consists of a per-unit length impedance *Z* (W/m) constituted by a right-handed per-unit-length inductance L_R (H/m) in series with a left-handed per-unit-length capacitance C_L (F/m), and a per-unit-length admittance *Y* (S/m) constituted by a right-handed per-unit-length capacitance C_R (F/m) in parallel with a left-handed per-unit-length inductance L_L (H/m). The complex propagation constant γ and the propagation constant β of the CRLH transmission line unit cell are given by

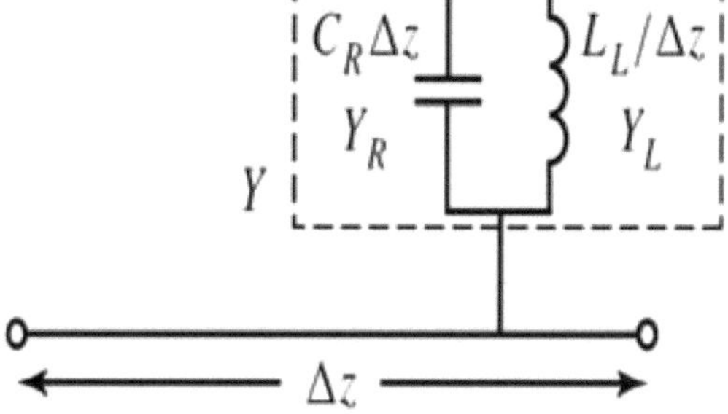

Figure 1. Equivalent circuit model for an ideal CRLH transmission line.

$$\gamma = \alpha + j\beta = \sqrt{ZY} \tag{1}$$

Where

$$\beta(\omega) = s(\omega)\sqrt{\omega^2 L_R C_R + \frac{1}{\omega^2 L_L C_L} - \left(\frac{L_R}{L_L} + \frac{C_R}{C_L}\right)} \tag{2}$$

$$s(\omega) = \begin{Bmatrix} -1\ if \omega < \omega_{se} = \min\left(\frac{1}{\sqrt{L_R C_L}}, \frac{1}{\sqrt{L_L C_R}}\right) \\ 0\ if \omega_{se} < \omega < \omega_{sh} \\ +1\ if \omega > \omega_{sh} = \max\left(\frac{1}{\sqrt{L_R C_L}}, \frac{1}{\sqrt{L_L C_R}}\right) \end{Bmatrix} \tag{3}$$

and

$$Z(\omega) = j\left(\omega L_R - \frac{1}{\omega C_L}\right) \tag{4}$$

$$Y(\omega) = j\left(\omega C_R - \frac{1}{\omega L_L}\right) \tag{5}$$

where $\beta(\omega)$, $s(\omega)$, $Z(\omega)$, and $Y(\omega)$ represent the dispersion relation, sign function, impedance, and admittance of the antenna structure, respectively. Series and shunt resonance frequencies, respectively, are given by

$$\omega_{se} = \frac{1}{\sqrt{L_R C_L}} \tag{6}$$

$$\omega_{sh} = \frac{1}{\sqrt{L_L C_R}} \tag{7}$$

The unit cell's permittivity and permeability are given by

$$\mu = \frac{Z}{j\omega} = L_R - \frac{1}{\omega^2 C_L} \tag{8}$$

$$\varepsilon = \frac{Y}{j\omega} = C_R - \frac{1}{\omega^2 L_L} \tag{9}$$

The prototype antenna based on CRLH transmission line unit cells, shown in **Figure 2**, consists of L- and T-shaped slits etched on a rectangular radiation patch. The patch is short-circuited to ground through high impedance microstrip lines that are coiled to reduce the antennas' size. Appropriate number of CRLH unit cells is cascaded together in the antenna that is terminated in a matched load to achieve the required bandwidth and radiation characteristics. The antenna was fabricated on glass-reinforced epoxy FR-4 substrate with a dielectric constant of 4.6, thickness of 0.8 mm, and loss tangent of 1×10^{-3}. Standard manufacturing technique was used to realize the antenna.

The antenna design was first analyzed and optimized using ANSYS high frequency structure simulator (HFSS™). Two waveguide ports were defined to represent the input and output of the antenna structure, as shown in **Figure 2(a)**, in order to evaluate its performance. The antenna was excited at port 1 through an SMA connector, and terminated to a matched 50 Ω load at port 2 using surface mount technology (SMD1206) of dimensions 3.5×1.8 mm^2.

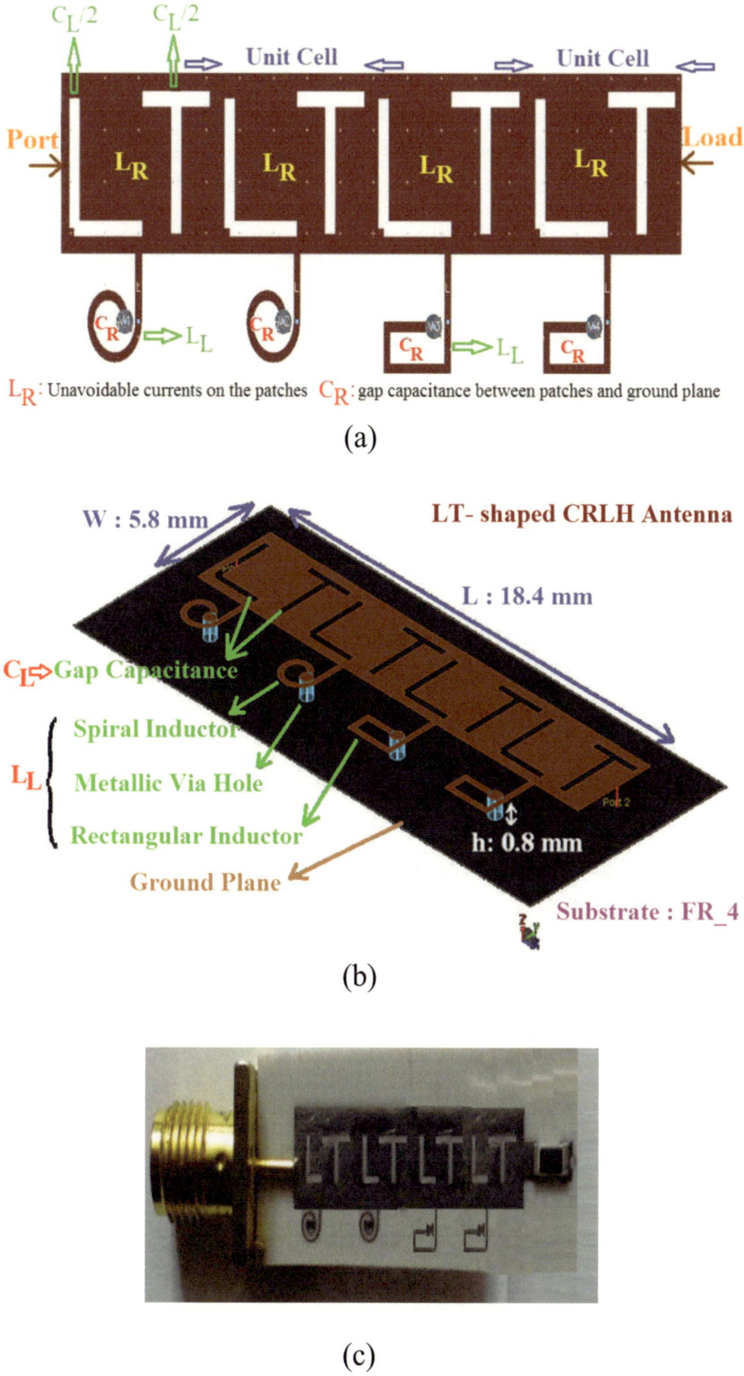

Figure 2. The proposed antenna based on four CRLH transmission line unit cells. (a) Configuartion of the proposed CRLH transmission line antenna. (b) Isometric view of the proposed CRLH transmission line antenna. (c) Fabricated antenna prototype.

A more accurate model of the proposed CRLH transmission line unit cell that is employed in the antenna is shown in **Figure 3**. The model includes loss components in the unit cell structure that are represented by right-handed resistance (R_R) and conductance (G_R), and left-handed resistance (R_L) and conductance (G_L). These parameters account for the radiation emitted by the antenna. Optimized magnitudes of these parameters in the unit cell were obtained from ANSYS HFSS™, that is, $L_L = 6$ nH, $C_L = 2.4$ pF, $L_R = 2$ nH, $C_R = 1$ pF, $R_L = 5\ \Omega$, $R_R = 3\ \Omega$, $G_L = 4.5$ ℧, and $G_R = 2$ ℧.

The actual antenna's dimensions are 22.6 × 5.8 × 0.8 mm³ or 0.037 λ_0 × 0.009 λ_0 × 0.001 λ_0, where λ_0 is free space wavelength at 0.5 GHz. The simulated and measured impedance bandwidth of the antenna are 11.1 GHz (0.35–11.45 GHz) and 10.8 GHz (0.5–11.3 GHz), respectively, for voltage standing wave ratio (VSWR) < 2. This corresponds to a fractional bandwidth of 188% for the simulation result, and 183% for the measured result. The reflection coefficients and the measured VSWR of the proposed miniaturized antenna are shown in **Figure 4**. The simulated and measured group delay response depicted in **Figure 5** shows the group delay variation is under 0.25 ns for a large frequency range up to 16 GHz.

Results of the parametric study are shown in **Figure 6**. It is evident from these results that the antenna's gain and radiation efficiency can be improved by increasing the number of CRLH unit cells. The peak gain and radiation efficiency are obtained at around 8 GHz. Gain of greater than 4 dBi from 2.9–12.6 GHz is achieved for a unit cell of four. Over this frequency range, the radiation efficiency exceeds 50%. As the proposed antenna had to fit within an area of 23 × 6 mm², the number of unit cells selected in the design was therefore four.

The simulated and measured gain and radiation efficiency of the antenna at various frequencies are given in **Tables 1** and **2**. The results in these tables show a better performance obtained from the antenna over higher frequencies than at the lower frequencies. The simulated three-dimensional (3-D) and measured two-dimensional (2-D) radiation patterns of the compact CRLH antenna at various frequencies are shown in **Figures 7** and **8**, respectively. The results show the cross-polarization is comparable or higher than copolarization at certain frequencies, however, over a small angular range in both the E- and H-planes. This is observed in

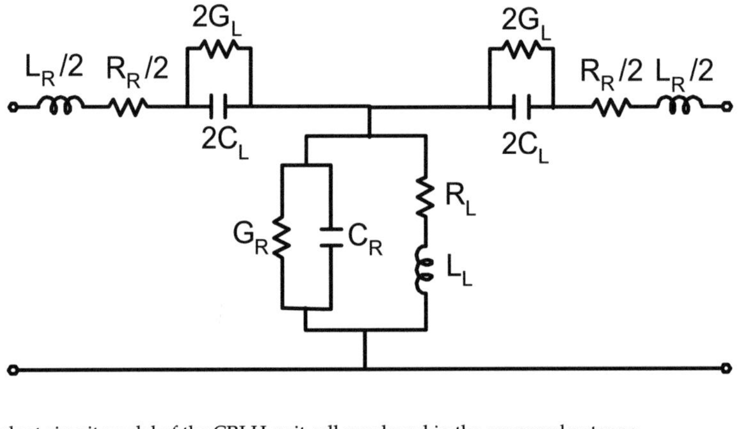

Figure 3. Equivalent circuit model of the CRLH unit cell employed in the proposed antenna.

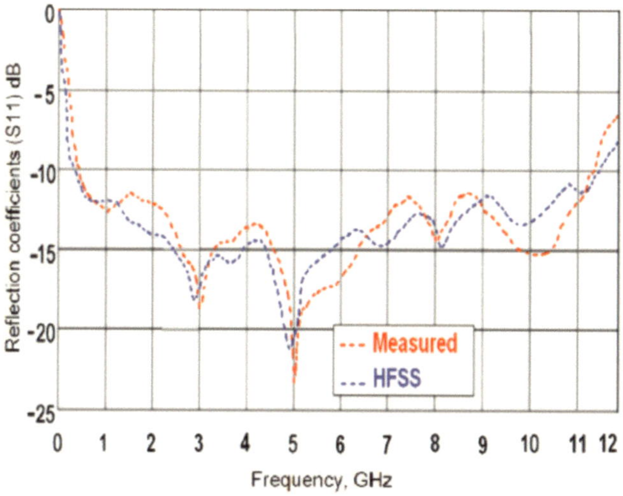

(a)

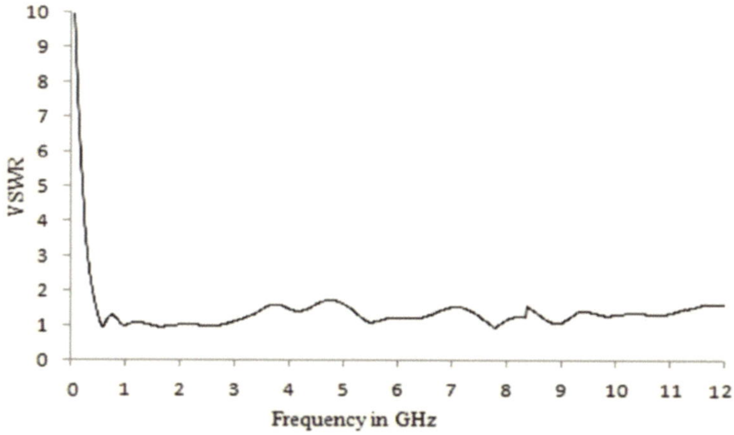

(b)

Figure 4. Reflection coefficient and measured VSWR response of the CRLH transmission line antenna. (a) Simulated and measured reflection coefficient response. (b) Antenna's measured VSWR response.

other CRLH antennas too [25–31] which present scope for improvement of such antennas. Cross-polarization can be reduced by loading the shorting pins, which has been demonstrated for patch antennas [32–35]. The shorting pins are located in the centerline of a square patch to strengthen the surface current density near the feeding point at edge. Because of symmetric arrangement of these two shorting pins, surface current density on the patch is maintained as the odd-symmetric property with respect to the H-plane, thus tremendously degrading the cross-polarization level. Such a solution is worthy of investigation in CRLH antennas.

The measured gain and radiation efficiency of the CRLH antenna is shown in **Figure 9**. The antenna has a peak gain of 6.5 dBi and radiation efficiency of 88% at around 8 GHz. The antenna has a gain that exceeds 4 dBi over 3.5–12.5 GHz, and radiation efficiency over this frequency range is greater than 53%.

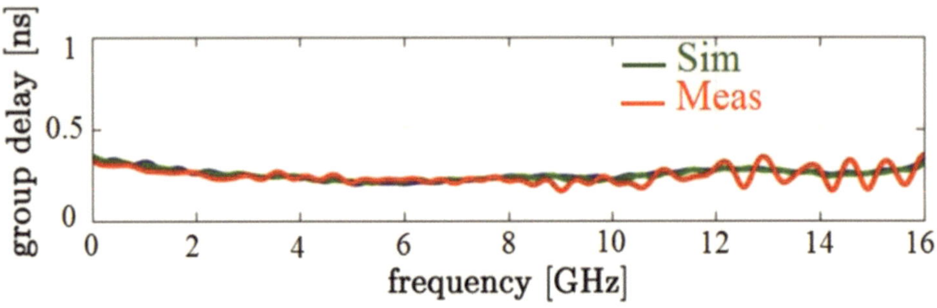

Figure 5. Group delay response of the CRLH transmission line antenna.

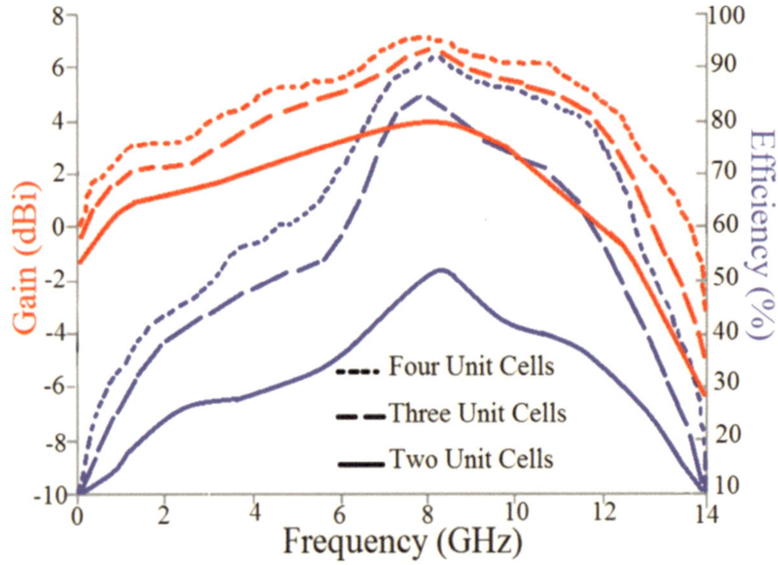

Figure 6. The antenna's simulated radiation gain and efficiency response as a function of number of unit cells.

Frequency (GHz)	0.5	3	5	8	11.3
Gain (dBi)	1.7	3.8	5.1	7	6.1
Efficiency (%)	25	50	62	91	79

Table 1. Simulated gain and radiation efficiency.

Table 3 shows the proposed antenna's salient features. The antenna is highly compact and operates over an ultra-wideband (UWB). In addition, it can be easily integrated with RF circuits making it suitable for UWB wireless communication systems. Characteristics of the CRLH antenna are compared with other recently reported antennas in **Table 4**.

http://dx.doi.org/10.5772/intechopen.68612

Frequency (GHz)	0.5	3	5	8	11.3
Gain (dBi)	1.5	3.4	4.8	6.5	5.7
Efficiency (%)	20	45	57	88	73

Table 2. Measured gain and radiation efficiency.

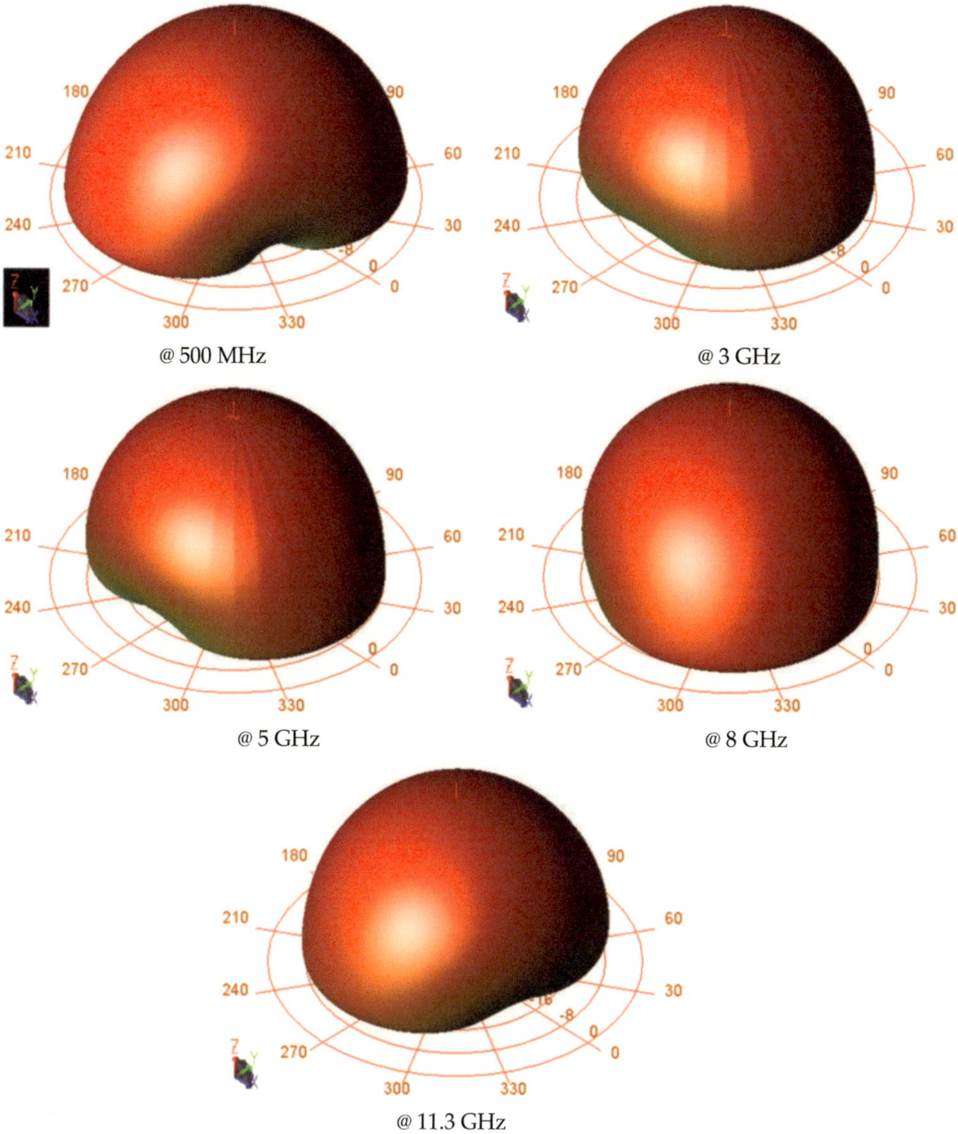

Figure 7. Simulated 3-D radiation patterns of the CRLH antenna.

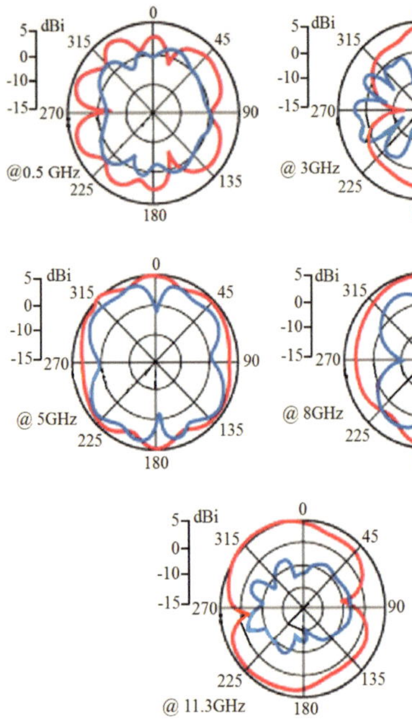

(a)

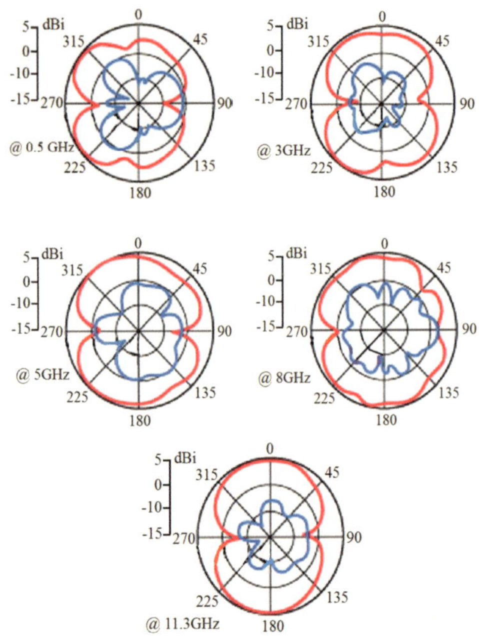

(b)

Figure 8. The measured radiation patterns of the CRLH transmission line antenna in the E- and H-planes. (a) Measured E-plane co and cross radiation patterns (Co: red line, Cross: blue line). (b) Measured H-plane co and cross radiation patterns (Co: red line, Cross: blue line).

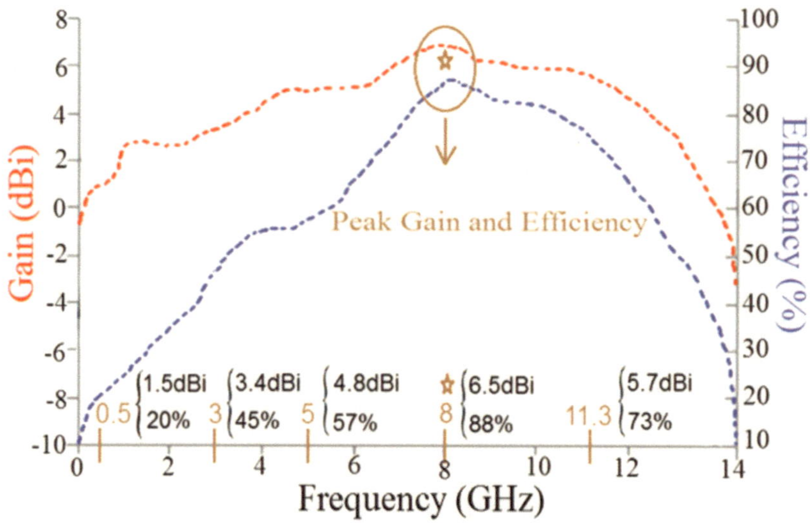

Figure 9. Measured gain and radiation efficiency response of the CRLH antenna.

Dimensions	22.6 × 5.8 × 0.8 mm^3
	0.037 λ_0 × 0.009 λ_0 × 0.001 λ_0 at 0.5 GHz
Bandwidth	10.8 GHz (500 MHz–11.3 GHz)
	(Fractional bandwidth = 183%)
Gain (dBi)	6.5 (max) at 8 GHz
	≥4 from 2.9–12.6 GHz
Efficiency (%)	88 (max) at 8 GHz
	≥50% from 2.9–12.6 GHz

Table 3. Measured characteristics of the CRLH antenna.

Reference	Dimensions	Fractional bandwidth (%)	Peak gain (dBi)	Max. efficiency (%)
[2]	0.051 λ_0 × 0.016 λ_0 × 0.002 λ_0	123.8	2.8	70
[6] with 7 unit cells	0.556 λ_0 × 0.179 λ_0 × 0.041 λ_0	87.16	3.4	68.1
[6] with 8 unit cells	0.564 λ_0 × 0.175 λ_0 × 0.02 λ_0	84.23	2.35	48.2
[23]	0.047 λ_0 × 0.021λ_0 × 0.002 λ_0	104.76	2.3	62
Proposed antenna	0.037 λ_0 × 0.009 λ_0 × 0.001 λ_0	183	6.5	88

Table 4. Comparison of the CRLH antenna with other reported antennas.

To summarize, this chapter presented the design and measured results of a novel antenna that is based on CRLH transmission lines. The antenna is highly compact planar structure with dimensions of 22.6 × 5.8 × 0.8 mm^3 and possesses desirable characteristics of ultra-wideband performance (500 MHz–11.3 GHz) with gain and radiation efficiency of 6.5 dBi and 88%, respectively, at 8 GHz. The low cost antenna is simple to design and easy to fabricate using standard manufacturing techniques. The CRLH transmission line unit cell constituting the antenna is realized by etching L- and T-shaped dielectric slots inside a rectangular patch, which is grounded through a high impedance transmission line. By cascading together several unit cells, the desired bandwidth and radiation characteristics can be obtained. In addition, the antenna can be easily integrated with RF electronics.

Author details

Mohammad Alibakhshikenari[1*], Mohammad Naser-Moghadasi[2], Ramazan Ali Sadeghzadeh[3], Bal Singh Virdee[4] and Ernesto Limiti[1]

*Address all correspondence to: alibakhshikenari@ing.uniroma2.it

1 Electronic Engineering Department, University of Rome "Tor Vergata", Rome, Italy

2 Faculty of Engineering, Science and Research Branch, Islamic Azad University, Tehran, Iran

3 Faculty of Electrical Engineering, K. N. Toosi University of Technology, Tehran, Iran

4 London Metropolitan University, Center for Communications Technology, London, UK

References

[1] Alibakhshi Kenari M. Design and modelling of New UWB metamaterial planar cavity antennas with shrinking of the physical size for modern transceivers. International Journal of Antennas and Propagation. 2013; **2013**:12. Article ID 56253. DOI: 10.1155/2013/562538.

[2] Anon., FCC First Report and Order on Ultra Wideband Technology, Feb. 2002.

[3] Wang YJ, Lee CK, Tian PS, Lee SW. "Novel microstrip-monopole integrated ultra wideband antenna for mobile UWB devices", Radio and Wireless Conference, 2003. RAWCON '03. Proceedings, pp. 87-90, 10-13 August 2003.

[4] Taniguchi T, Kobayashi T. An Omnidirectional and low-VSWR antenna for the FCC-approved UWB frequency band. IEEE International Symposium on Antennas and Propagation. 2003;**13**:460-463

[5] Sorgel W, Waldschmidt C, Wiesbeck W. Transient response of a vivaldi antenna and a logarithmic periodic dipole array for ultra wideband communication. IEEE International Symposium on Antennas and Propagation. 2003;**3**:592-595

[6] Alibakhshi-Kenari M, Naser-Moghadasi M, and Sadeghzadeh RA. Composite right-/left-handed-based antenna with wide applications in very-high frequency/ultra-high

frequency bands for radio transceivers. IET Microwaves, Antennas & Propagation. 2015;**9**(15):1713 –1726.

[7] Liang J, Chiau CC, Chen X, Parini CG. Study of a printed circular disc monopole antenna for UWB systems. IEEE Transactions of Antennas Propagation. 2005;**53**(11):3500-3504

[8] Chen ZN, See TSP, Qing X. Small printed ultrawideband antenna with reduced ground plane effect. IEEE Transactions of Antennas Propagation. 2007;**55**(2):383-388

[9] Dissanayake T, Esselle KP. Prediction of the notch frequency of slot loaded printed UWB antennas. IEEE Transactions of Antennas Propagation. 2007;**55**(11):3320-3325

[10] Wheeler HA. Fundamental limitations of small antennas. Proceedings of IRE.1947; **35**:1479-1484

[11] Caloz C, Itoh CT. Electromagnetic Metamaterials Transmission Line Theory and Microwave Applications, New Jersey: John Wiley & Sons, Inc.; 2006

[12] Porath R. Theory of miniaturized shorting-post microstrip antennas. IEEE Transactions of Antennas Propagation. 2000;**48**(1):41-47

[13] Stutzman WL. Antenna Theory and Design. J. Wiley & Sons.; 1997

[14] Balanis CA. Antenna Theory and Design. John Wiley & Sons.; 2016

[15] Jahani S, Rashed-Mohassel J, Shahabadi M. Miniaturization of circular patch antennas using MNG metamaterials. IEEE Antennas and Wireless Propagation Letters. 2010;**9**:1194-1196

[16] Wang C, Hu BJ, Zhang XY. Compact triband patch antenna with large scale of frequency ration using CRLH-TL structures. IEEE Antennas and Wireless Propagation Letters. 2010;**9**:744-746

[17] Bilotti F, Alu A, Vegni L. Design of miniaturized metamaterial patch antennas with μ-negative loading. IEEE Transaction on Antennas and Propagation. 2008;**56**(6):814-818

[18] Engheta N, Ziolkowski RW. Metamaterials: Physics and Engineering Explorations. New York: Wiley; 2006

[19] Eleftheriades GV, Balmain KG. Negative Refraction Metamaterials: Fundamental Principles and Applications. New York: John Wiley & Sons; 2005

[20] Slyusar VI. "Metamaterials on antenna solutions", International Conference on Antenna Theory and Techniques (ICATT). 2009.

[21] Lee CJ, Leong K, Itoh T. Composite right/left-handed transmission line based compact resonant antennas for RF module integration. IEEE Transactions of Antennas and Propagation. 2006;**54**(8):2283-2291

[22] Lee CJ, Huang W, Gummalla A, Achour M. Small antennas based on CRLH structures: Concept, design, and applications. IEEE Antennas and Propagation Magazine. 2011;**53**(2):10-25

[23] Alibakhshi Kenari M. Printed planar patch antennas based on metamaterial. International Journal of Electronics Letters. 2014;**2**(1):37-42. http://dx.doi.org/10.1080/21681724.2013.874042

[24] Alibakhshi-Kenari M, Movahhedi M, Naderian H. A new miniature ultra wide band planar microstrip antenna based on the metamaterial transmission line. IEEE Asia-Pacific Conference on Applied Electromagnetics (APCAE), Dec. 2012, pp. 293-297, Malaysia.

[25] Zheng L, Quan XL, Liu HJ, Li RL. Broadband planar antenna based on CRLH structure for DVB-H and GSM-900 applications. Electronics Letters. 2012;**48**(23):1443-1444

[26] Wu GC, Wang GM, Peng HX, Gao XJ, Liang JG. Design of leaky-wave antenna with wide beam-scanning angle and low cross polarisation using novel miniaturised composite right/left-handed transmission line. IET Microwave Antennas Propagation. 2016;**10**(7):777-783

[27] Nandi S, Mohan A. A miniaturized dual mode CRLH unit cell loaded SIW antenna. 5th IEEE Applied Electromagnetics Conference (AEMC) 2015, 18th- 21st, December 2015, pp. 1-2.

[28] Lee H, Woo DJ, Nam S. Compact and bandwidth-enhanced asymmetric coplanar waveguide (ACPW) antenna using CRLH-TL and modified ground plane. IEEE Antennas and Wireless Propagation Letters. 2016;**15**:810-813

[29] Saurav K, Sarkar D, Srivastava KV. CRLH unit-cell loaded multiband printed dipole antenna. IEEE Antennas and Wireless Propagation Letters. 2014;**13**:852-855

[30] Yan S, Soh PJ, Vandenbosch GAE. Wearable dual-band composite right/left-handed waveguide textile antenna for WLAN applications. Electronics Letters. 2014;**50**(6):424-426

[31] Xu HX, Wang GM, Qi MQ, Cai T. Compact fractal left-handed structures for improved cross-polarization radiation pattern. IEEE Transactions on Antennas and Propagation. 2014;**62**(2):546-554

[32] Zhang X, Zhu L. Patch antennas with loading of a pair of shorting pins toward flexible impedance matching and low cross polarization. IEEE Transactions on Antennas and Propagation. 2016;**64**(4):1226-1233

[33] Ghosh D, et al. Physical and quatitative analysis of compact rectangular microstrip antenna with shorted non radiating edges for reduced cross polarized radiation using modified cavity model. IEEE Antennas & Propagations Magazine. 2014;**56**(4):61-72.

[34] Ghosh A, et al. Rectangular microstrip antenna on slot-type defected ground for reduced cross-polarized radiation. IEEE Antennas and Wireless Propagation Letters. 2015;**14**:321-324

[35] Chakraborty S, et al. Substrate fields modulation with defected ground structure: A key to realize high gain, wideband microstrip antenna with improved polarization purity in principal and diagonal planes. International Journal of RF and Microwave Computer Aided Engineering, (Wiley). 2016;**26**(2):174-181

Section 2

Modern Approach to Microstrip Antenna Analysis

Chapter 3

Design and Analysis of Microstrip Patch Antennas Using Artificial Neural Network

Vivek Singh Kushwah and Geetam Singh Tomar

Additional information is available at the end of the chapter

http://dx.doi.org/10.5772/intechopen.69522

Abstract

The microstrip patch antenna can also be designed using an artificial neural network (ANN) modeling technique where size of the antenna is major limitation especially in mobile and wireless applications. In this chapter, analysis and synthesis problems for designing of microstrip patch antennas were discussed using the artificial neural network technique. An analysis problem refers to calculation of resonant frequency of microstrip patch antenna whereas a synthesis problem refers to calculation of dimensions of patch antenna. Both problems are reciprocal of each other. Results are implemented using graphical user interface (GUI) tools of MATLAB programming language. Back-propagation training algorithm of artificial neural network is used to train the network for minimization of error and computation time. Therefore, the geometric dimensions of patch are obtained with high accuracy in less computation time as compared to simulation software.

Keywords: microstrip antennas, artificial neural networks, ANN modeling, IE3D electromagnetic simulator, resonant frequency, return-loss, GUI

1. Introduction

Nowadays, many antennas are used for various wireless applications but microstrip antenna is the most preferable antenna for microwave communication. The microstrip antennas are low cost and very easily used fabrication techniques [1, 2]. Since a photoetching process is applied for the fabrication of microstrip patch antennas hence it is also known as "patch antennas." The shape of this patch may be square, hexagonal, rectangular, trapezoidal, or any other pattern. In this chapter, rectangular microstrip patch antennas are designed for synthesis and analysis by an artificial neural network (ANN) modeling procedure.

There has been a tremendous development in the field of patch antenna during the last two decade [3, 4]. It is the result of the overwhelming research and this technology has been quickly absorbed for consumption. A few previous research works done in this field are discussed in this section. The length of an antenna is inversely proportional to the resonant frequency of patch, that is, an antenna operating at higher frequency is smaller in size than compared to the antenna operating at lower frequency. So, to decrease the antenna dimensions required for operating the antenna at lower frequency, the electrical length of the patch should be enlarged at lower frequency. This technique has been successfully used for minimizing the antenna dimensions. A technique of improving the electrical length is by meandering the surface current path of the patch. One such research work was done by Dey and Mittra [5]. In the mentioned article, 70% size reduction of the proposed microstrip patch design is reported. In Ref. [6], authors used three meandering slots and a shorting pin on a rectangular patch to obtain the results. In Ref. [7], an innovative technique is proposed for calculating the resonant frequency of circular microstrip patch antennas using artificial neural for computer-aided design (CAD) applications. In Ref. [8], the back-propagation multilayered-perceptron network is used for calculating the resonant frequency of electrically thin and thick rectangular microstrip antennas. This technique is applicable for a wide range of substrate thicknesses and permittivity for the computer-aided design (CAD) applications of microstrip antennas. One such research work was conducted by Kuo et al. [9]. In this work, the researchers have chosen the ground plane for modification in designed antenna. A total number of three slots were loaded in the ground plane. The authors have shown that the variation seen by changing the length of the slots. For the best case, the size reduction of 56% was obtained by the authors. Some researchers also reported size reduction using inductive loading. One such research work was reported by Reed et al. [10]. In the design, the researchers have loaded a printed inductor to minimize the frequency of resonant antenna. Using this, a size reduction of 50% was achieved. A bow-tie patch antenna loaded with Sierpinski fractal was also proposed by Anguera et al. [11] which has good directivity along with compactness of 42%. One of the methods of increasing the electrical path length is by meandering the surface current path of the patch. In [13], a new method for developing computer-aided design (CAD) models for microstrip antenna is proposed using spectral domain (SD) formulation and artificial neural network (ANN) technique. In Ref. [14], design of rectangular microstrip antennas is proposed using the artificial neural network technique for the most efficient dielectric materials. In Ref. [15], the authors have suggested the combination of inductive loading and shorting pin. The authors have suggested that using the novel design, a size reduction of 75% may be achieved. A bow-tie patch antenna loaded with Sierpinski fractal was also presented by Garima et al. [16] along with compactness of 42% and good directivity. In Ref. [17], single-fed broadband square patch antenna was constructed for ultra high frequency (UHF) radio frequency identification (RFID) applications using meandered probe feeding techniques for good impedance matching. The purpose of the antenna was to optimize and modify antenna. In Ref. [18], two L-shaped patch antenna was constructed using a probe or coaxial feeding techniques. Two parallel L-shaped antennas were used to obtain better bandwidth. The bandwidth of the antenna was 6.41 GHz ranging from 3.51

to 9.65 GHz. Return-Loss of the proposed antenna is −28 dB at 3.51 GHz and −25 dB at 9.65 GHz. In [19], E-shaped antenna was constructed for wireless communications, and its applications using U-slot patch was introduced under patch antenna to increase and optimize bandwidth of the proposed antenna. The systematic study of antenna was enhanced gain, return loss, bandwidth, and radiation pattern. In Ref. [20], four mini fractal antennas were constructed to improve bandwidth and size of the antenna using coaxial feeding techniques. The proposed design was used for broadband applications. Four spring resonators form an ultra-wideband frequency. Low percentage error and high precision in less time are obtained by using a knowledge-based hybrid neural network (KBHNN) model for designing of microstrip antennas (proximity coupled), which is used for various WiMAX, WiFi, and WLAN applications. **Figure 1** illustrates a simple rectangular microstrip patch antenna of substrate height h, dielectric constants ε_r, ε_y, width W, and length L.

In this chapter, synthesis is defined as to obtain dimensions (W, L) of microstrip antenna while providing the resonant frequency (f_r), height of the dielectric substrate (h), and dielectric constants at the input of the ANN model (**Figure 2**). Dimensions of patch are computed using the designed equations of the microstrip patch antennas. For the analysis problem of patch antenna, resonant frequency (f_r) or both upper and lower cutoff frequencies are obtained at the output of ANN model while providing the dimensions of patch and other parameters as the inputs of ANN model, as shown in **Figure 3**. This model is very significant for antenna researchers to determine the dimensions and other parameters of microstrip antenna.

The range of dielectric constants should be taken between 2.2 and 12 but the dielectric substrate should be thicker and the dielectric constant should be less for obtaining high efficiency and

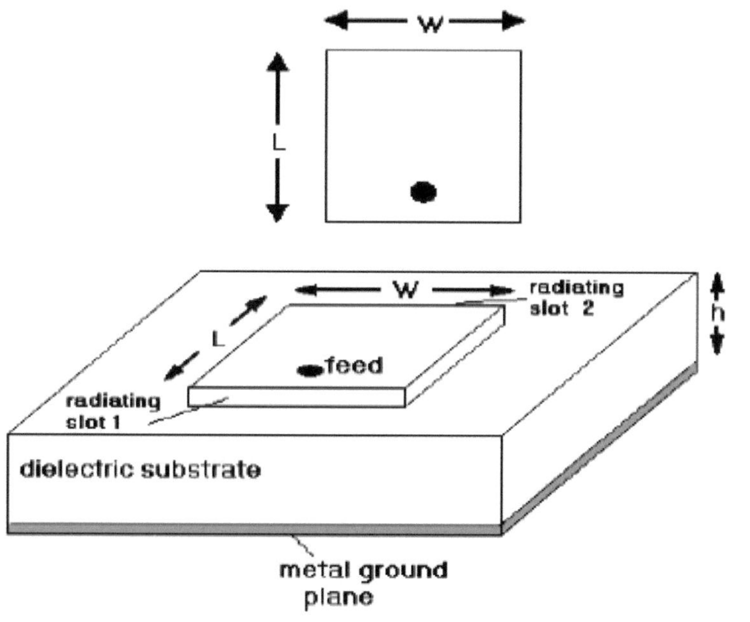

Figure 1. Basic layout of rectangular patch antenna.

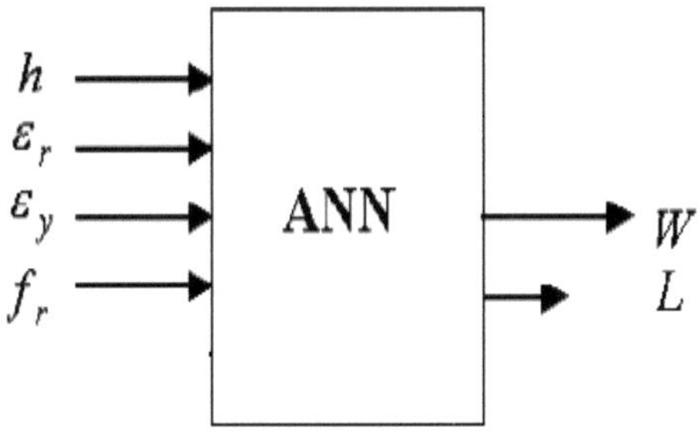

Figure 2. The synthesis of microstrip patch using ANN model.

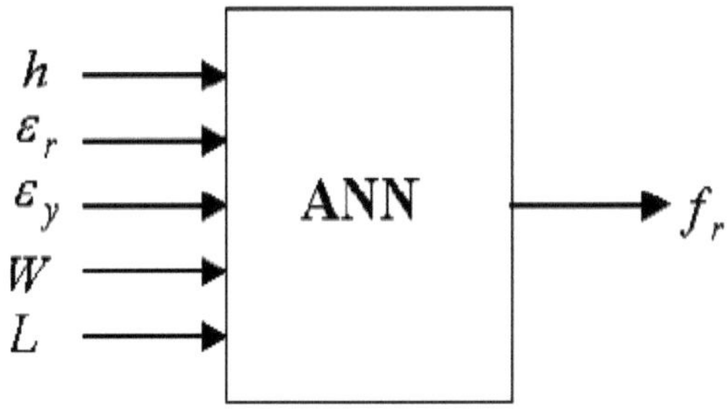

Figure 3. The analysis of microstrip patch using ANN technique.

wide bandwidth. The effective thickness (h_e) and effective dielectric constant ($\varepsilon_{\text{reff}}$) are calculated using Eqs. (1) and (2), where h is thickness and ε_g is the geometric mean of dielectric constant.

$$h_e = \sqrt{\frac{\varepsilon_r}{\varepsilon_g}}h \tag{1}$$

$$\varepsilon_{eff} = \frac{\varepsilon_g + 1}{2} + \frac{\varepsilon_g - 1}{2}\left[1 + 12\frac{h_e}{W}\right]^{-\frac{1}{2}} \tag{2}$$

If the velocity of light is denoted by C, the width of antenna is calculated using Eq. (3).

$$W = \frac{C}{2f_o}\sqrt{\frac{2}{\varepsilon_g + 1}} \tag{3}$$

Eqs. (4) and (5) represents the real length of the patch (L).

$$L = L_{eff} - 2\Delta L \tag{4}$$

$$L = \frac{1}{2f_r\sqrt{\varepsilon_{eff}}\sqrt{\mu_o\varepsilon_r}} - 2\Delta L \tag{5}$$

The fringing property of patch gives rise to the extended electric length, which is illustrated by Eq. (6)

$$\Delta L = 0.412h \frac{(\varepsilon_{reff} + 0.3)\left(\frac{w}{h} + 0.264\right)}{(\varepsilon_{reff} - 0.258)\left(\frac{w}{h} + 0.813\right)} \tag{6}$$

2. Synthesis and analysis problem for microstrip patch antenna

Synthesis is defined as to obtain the dimensions of microstrip patch while providing the resonant frequency, thickness, and dielectric constants of the dielectric material as the input parameters of ANN model (**Figure 2**). Dimensions of patch are computed using the designed equations of the microstrip patch antennas. For the analysis problem of patch antenna, resonant frequency (f_r) or both upper and lower cutoff frequencies are obtained at the output side of ANN model while providing the dimensions of patch (W, L) and other parameters (ε_r, ε_y, h) at the input side of ANN model (**Figure 3**). These ANN models are easily applicable and very useful for researchers to judge the accurate upper and lower cutoff frequency, bandwidth, resonant frequency, and accurate dimensions of patch antenna. Here ε_r and ε_y represent the electrical properties of the dielectric material.

2.1. ANN architecture for microstrip antenna

There are many algorithms of ANN which is used to train the neural network [12]. In this work, back-propagation training algorithm was applied to train the neural network and to build the ANN models for patch antenna. This algorithm is explained in Section 2.1.1.

2.1.1. Back-propagation algorithm

In this algorithm, connections are done from input layer to hidden layer and to output layer. The network is trained using the gradient descent method and the error calculated is sent back to the hidden layer as well as the output layer for weight adjustments [13]. In multilayer perceptron network there are two sets of weights, namely from input layer to hidden layer and from hidden to output layer. The error due to the second set of weights is calculated using the delta rule and the error is required to propagate from output layer to input layer so that the error can be assigned proportionately to the weights which caused it. This problem is known as credit-assignment problem and helps in deciding which weights should be altered and by how much. This problem can be solved by using the error back-propagation algorithm, which works in two passes:

1. **Forward pass**: it takes input vector, computes function, and evaluates derivative of error function with respect to weights.

2. **Backward pass**: it propagates the error derivatives backward and computes the weight adjustments.

This algorithm uses the supervised learning mechanism.

In multilayer feed-forward networks, normally, there is an input layer, one or two hidden layers and one output layer. The raw values are given to the input layer, computations are done at the hidden layer and their activations are given to the output layer where each pattern is given a specific classification category. In this algorithm, we are giving a solution to credit assignment problem using partial derivatives to find out how weight w_{ij} effects the error which is represented by Eq. (7).

$$\frac{\partial E(t)}{\partial W_{ij}(t)} \tag{7}$$

Here, partial derivatives are used because the direction of the errors can be determined, which may be positive or negative. If this derivative to be negative, this will add in weight and finally decreases the error and thus local minima can be reached (see step 6 of algorithm). This shows that if derivative of error is positive, the weight increases by adding a negative value to the weight and conversely, if negative. Procedure of taking derivative in this algorithm starts from weights of output layer to input layer, that is why, it is known as "back-propagation algorithm." **Figure 4** represents the three layers fully connected feed-forward network using the back propagation algorithm [14]. It consists of three layers: one input layer, one hidden layer, and one output layer. Here, $x_1, x_2, \ldots, x_n$ are the input neurons. x_b is the bias input, with associated weight w_{1b}. w_{1b} can increase or decrease in the net output of activation function depending on its positive or negative value.

The value of w_{1b} is kept constant throughout the process. It can be 0 or $\pm$ 1. $y_1, y_2, \ldots, y_j$ are the hidden neurons and $O_1, O_2, \ldots, O_m$ are the output neurons. v_{jn} represents the input hidden layer weights, that is, these are the weights associated with input layer neurons connected to hidden layer neurons. w_{mj} represents the output-hidden layer weights, that is, these are the weights associated with hidden layer neurons connected to output layer. Input, hidden, and output layers can be represented with the help of vectors as illustrated in Eqs. (8-10).

$$x = [x_1, x_2, \ldots, x_n]^t \leftarrow \text{input training vector} \tag{8}$$

$$y = \left[y_1, y_2, \ldots, y_j\right]^t \leftarrow \text{hidden layer vector} \tag{9}$$

$$O = [O_1, O_2, \ldots, O_m]^t \leftarrow \text{output vector} \tag{10}$$

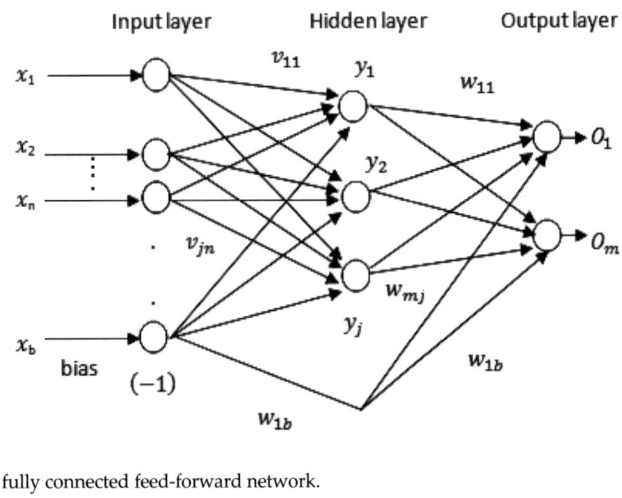

Figure 4. Three-layer fully connected feed-forward network.

where 't' indicates transposition of matrix.

Given are P training pairs

$\{x_1, a_1, x_2, a_2,\ldots, x_P, a_P\}$ and there is a bias $= -1$, η is the learning constant. Here a_1, a_2, and a_P are the desired output vector response.

Step 1: $\eta > 0$, Maximum error $E_{\max}$ is chosen.

Step 2: Weights are initialized at small random values, $E\leftarrow 0, p\leftarrow 1, k\leftarrow 1$

where E = present error; p = counter with in the training cycle; k = training step counter.

Step 3: Now begin the training process. Input is provided to input layer and output is Calculated as given in Eqs. (11) and (12).

$$x \leftarrow x_P, a \leftarrow a_P$$

$$y_j = f\left(v_j^t x\right) \text{ where } j = 1, 2, 3, 4, \ldots, J \tag{11}$$

$$O_k = f\left(w_j^t y\right) \text{ where } k = 1, 2, 3, 4, \ldots, K \tag{12}$$

Here f (.) is the activation function.

Step 4: Calculate the new value of error by using previous value of error, desired output a_k and actual output O_k using Eq. (13).

$$E' = \frac{1}{2}(a_k - O_k)^2 + E \text{ where } k = 1,\ 2,\ 3,\ 4, \ldots, K \tag{13}$$

Step 5: Now value of error for both the layer is computed.

Error signal for output layer is calculated using Eq. (14).

$$\delta_{ok} = \frac{1}{2}(a_k - O_k)(1 - O_k{}^2) \text{ for } k = 1,\ 2,\ 3,\ 4, \ldots, K \tag{14}$$

Eq. (15) represents the Value of error for middle (hidden) layer

$$\delta_{yj} = \frac{1}{2}\left(1 - y_j\right)^2 \sum_{k=1}^{k} \delta_{ok}\, w_{kj} \text{ for } j = 1,\ 2,\ 3,\ 4, \ldots, J \tag{15}$$

Step 6: Now weights of output layer are adjusted using Eq. (16):

$$w_{kj}' = w_{kj} + \eta\delta_{ok}y_j \text{ where } k = 1,\ 2,\ 3,\ 4, \ldots, K, \text{ where } j = 1,\ 2,\ 3,\ 4, \ldots, J \tag{16}$$

Step 7: Now weights of Hidden layer are modified using Eq. (17).

$$v_{ji}' = v_{ji} + \eta\delta_{yj}x_i \text{ for } i = 1,\ 2,\ 3, \ldots, I \tag{17}$$

Step 8: If p is less than P then $p = p + 1$ and $k = k + 1$ and go to step 3, otherwise return back to step 9.

Step 9: At last, the sequence of full training process is finished.

If the value of present error (E) < maximum error (E_{max}), then stop the training process and now check out weights of output layer w and weights of hidden layer v.

If the present value of error (E) > maximum error (E_{max}), then $E = 0$, $p = 1$ and start a new training sequence by return back to step 3.

In first step of training, η and E_{max} are chosen. Step 2 is the initialization step where values of p, k, E, and weights are initialized. Step 3 is feed-forward step where both hidden layer and output layer outputs are calculated. In step 4, errors are computed. In step 5, back-propagation training takes place and error signal is computed for both hidden layer and output layer. In step 7, output layer and hidden layer weights are adjusted, respectively. In step 8, it is checked if $p < P$, then increment p and k and go to step 3, otherwise go to step 9. In step 9, if $E < E_{max}$, the training is complete and if $E > E_{max}$, a new training cycle is started.

2.1.2. *Learning factors of back-propagation network*

The training of back-propagation networks is dependent on various parameters such as initial weights, number of layers, number of neurons per layer, and updation rule.

1. **Initial weight**: the choice of initial weights determines how fast the networks will converge. Typical value for choosing initial weights is between −1.00 and 1.00 or −0.5 and 0.5. The final solution may be affected by the initial weights of multilayer feed-forward networks. One method of choosing the weight w_{ij} choosing is in the range $\left[\frac{-3}{\sqrt{O_i}}, \frac{3}{\sqrt{O_i}}\right]$, where O_i is the number of processing elements.

   If all the weights are given equal values, the network may not be trained properly. If large values are given to weights at the initial stage, then the system may be stuck at local minima very near to starting point itself. So, it is necessary to initialize with small weights uniformly distributed in a small range

2. **Learning rate (η)**: the learning rate affects the convergence of back-propagation. A large value of η simply increases the speed of convergence but this large value introduces instability into the learning rule causing oscillations in the learned weights and on the other hand, if η has small value, the speed of training will be less. For successful operation, the range of η lies between10^{-3} and 10.

3. **Number of training data**: the numbers of input nodes are determined by the dimension, or size or the input vector to be classified or generalized with a certain output quality. The training data should be sufficient and proper for training.

4. **Momentum:** the gradient decent is very slow using small η and oscillates using large η. The problem of the oscillations can be sorted out by adding a momentum factor to the weight adjustments as illustrated in Eq. (18)

http://dx.doi.org/10.5772/intechopen.69522

$$\Delta w(t) = -\eta \nabla E(t) + \alpha \Delta w(t-1) \tag{18}$$

where α is the selected momentum constant having values between 0.1 and 0.8. $\alpha \Delta w(t-1)$ is called the momentum term. t and $(t-1)$ indicates the current and most recent training step, respectively.

5. **Number of hidden layers nodes**: in the case of all multilayer feed-forward networks, the size of hidden layer is determined experimentally. For a network of a reasonable size, the size of a hidden layer has to be only a small fraction of the input layer. For example, if the network does not converge to a solution, it may need more hidden layers. On the other hand, the user may use a small number of hidden layers when the network converges.

6. **Stopping criteria**: stopping criteria are used to decide when the network problem has been solved. It is possible to stop when:

   i. The mean squared error is sufficiently small.

   ii. The rate of change of the mean squared error is sufficiently small.

   iii. Combination of above two criteria.

2.1.3. *Realization of patch antenna with numerical results*

IE3D electromagnetic simulation software is used for designing of the rectangular patch antenna using the following design parameters.

Resonant frequency = 4.99 GHz, dielectric constant = 4.4, height of dielectric material = 1.6 mm

Now, the dimensions of patch can be calculated using Eqs. (1)–(6) for 4.99 GHz resonant frequency and given by

$L = 13.34$ mm, $W = 17.7$ mm

The basic design of microstrip patch antenna is shown in **Figure 5**. Here, 50 ohm port (Z_o) is connected to the patch by using the microstrip feed line technique. Feed is the point where patch radiates maximum. Eq. (19) is used to calculate the width W_o of microstrip line, where $\varepsilon_{\mathrm{reff}}$ is the effective dielectric constant.

$$Z_0 = 120\Pi/[\surd(\varepsilon_{reff})\{W_0/h + 1.393 + .667\ln(W_0/h + 1.444)\}] \tag{19}$$

which gives $W_0 = 2.94$ mm.

When IE3D electromagnetic simulation is performed, then we obtain the different resultant graph between return loss and frequency. While changing the dimension of patch, the different resultant graphs are obtained between return loss and frequency using simulation software, as shown in **Figures 6–8**. Here all the dimensions are taken in mm.

(1) When $W = 17.7$, $L = 13.34$, $\varepsilon_r = 4.4$, $h = 1.6$

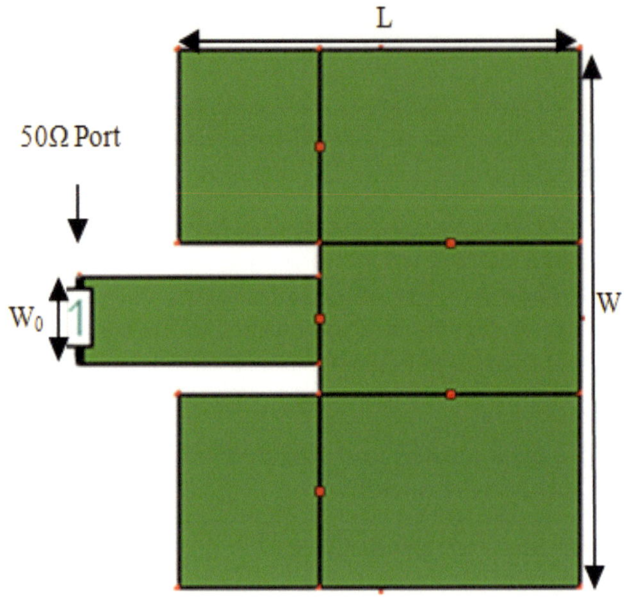

Figure 5. Design of microstrip patch antenna using simulation software.

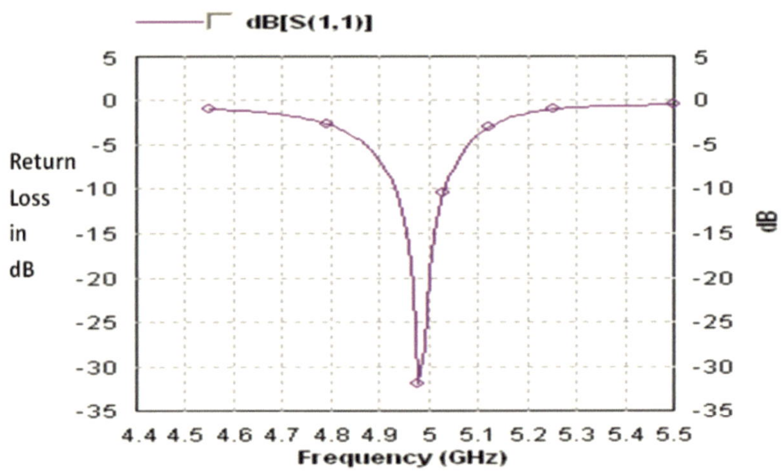

Figure 6. Simulated response of patch at 4.99 GHz resonant frequency.

Here, 4.94 GHz is the lower cutoff frequency (f_1) and 5.04 GHz is the upper cutoff frequency (f_2). Furthermore, only one dimension of the patch will be varied while maintaining all other parameters constant and vice versa. Frequency response between return loss and frequency are obtained, as shown in **Figure 7**, which gives the upper and lower cutoff frequencies (f_2, f_1) when the width and length of patch is 17.7 and 13.55 mm, respectively.

(2) When $W = 17.7$ mm, $\varepsilon_r = 4.4$, $L = 13.55$ mm, $h = 1.6$ mm

(3) When $W = 17.7$ mm, $L = 14.15$ mm, $h = 1.6$, $\varepsilon_r = 4.4$ mm

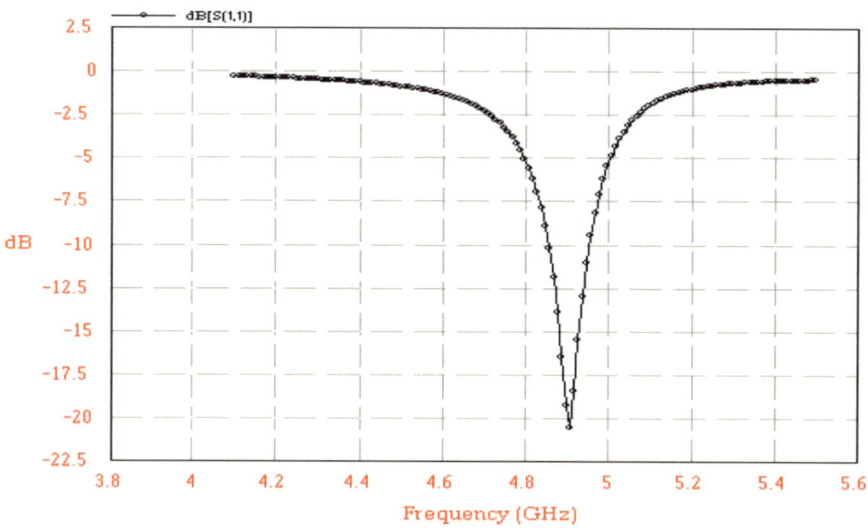

Figure 7. Simulated response of patch at 4.89 GHz resonant frequency.

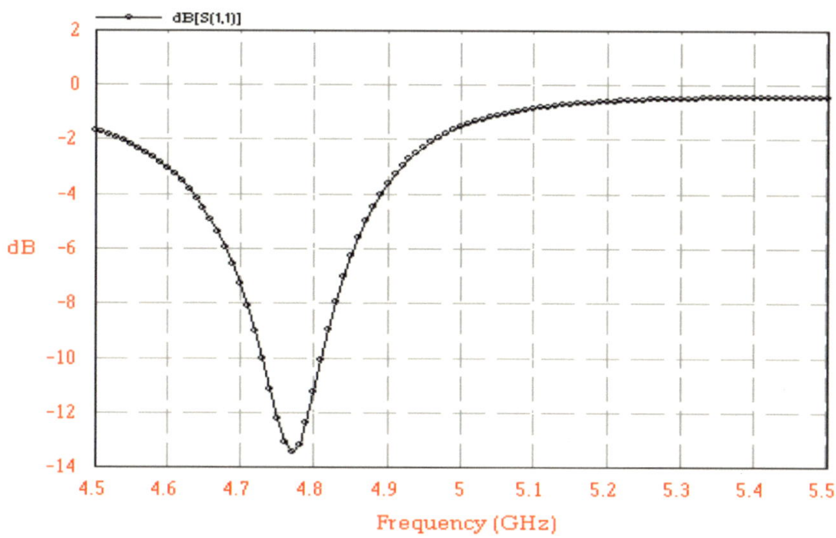

Figure 8. Simulated response of patch at 4.77 GHz resonant frequency.

Figures 6–8 represent the variation of frequency range of microstrip patch antenna with respect to variation in dimensions. Around 25–30 samples are collected using the EM simulation tool for ANN training for different dimensions, which are illustrated in **Table 1**. It is used for the analysis of patch antenna.

Now, the neural network is constructed for rectangular microstrip patch antenna. For the analysis problem of neural network, the dimensions (W, L) are considered as the input parameters of the ANN model whereas output parameters are upper and lower cutoff frequencies,

Inputs (dimensions of patch in mm)		Targets (lower and upper cutoff frequency in GHz)	
W	L	f_1	f_2
17.7	13.34	4.94	5.04
17.7	13.55	4.86	4.95
17.7	13.65	4.82	4.91
17.7	13.85	4.8	4.89
17.7	14.05	4.77	4.85
17.7	14.15	4.73	4.81
17.7	14.25	4.71	4.79
17.7	14.35	4.69	4.76
17.7	14.45	4.66	4.73
18.3	13.85	4.78	4.87
18.3	14.35	4.65	4.73
18.8	14.35	4.61	4.7
18.8	14.85	4.49	4.55
19.3	14.85	4.47	4.55
19.3	15.35	4.37	4.41
19.8	15.35	4.35	4.41
20.3	15.85	4.31	4.37
20.3	16.35	4.29	4.35
20.8	16.35	4.27	4.33
20.8	16.85	4.26	4.33
21.3	16.85	4.21	4.27
21.3	17.35	4.16	4.21
21.8	17.35	4.14	4.19
21.8	17.85	4.02	4.04
22.3	17.85	3.99	4.01
22.3	18.35	3.92	3.93

Table 1. Simulated results for ANN analysis of patch.

which act as targets. The artificial neural network is trained by using the ANN feed-forward back-propagation training algorithm and transfer function so that error can be minimized with less computation time. Matlab neural network graphical user interface (GUI) toolbox is used as a plateform for ANN training. The "***nntool***" command is used for ANN training in the GUI, which will display the network or data manager window, as given in **Figure 9**.

After displaying this window, input and outputs are provided in the matrix form by clicking on "**new data.**". Now, a neural network is created by clicking on "**New network.**" Here, the feed-forward back-propagation algorithm and trainlm (Levenberg-Marquardt back-propagation

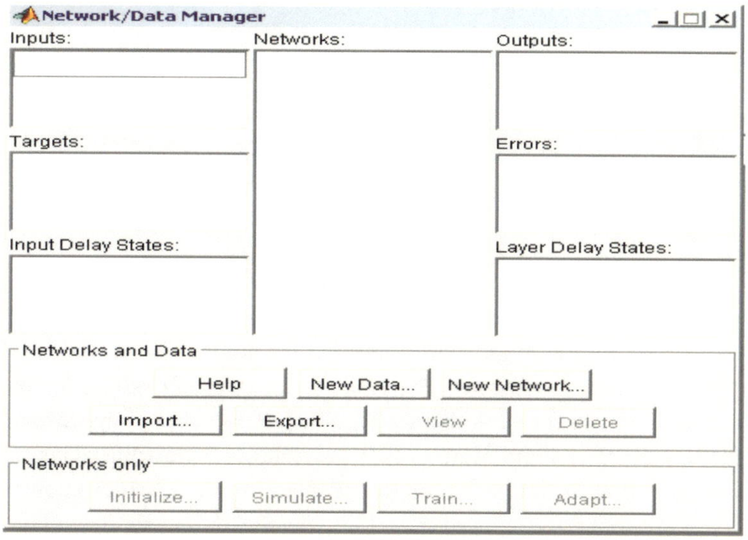

Figure 9. GUI of ANN.

training function) are applying where supported transfer functions are **purelin** (linear transfer function) and **tansig** (hyperbolic tangent sigmoid transfer function).

2.1.4. *Analysis of microstrip patch antenna using artificial neural network*

The network architecture is one of the basic building blocks for ANN to be deployed. It determines how the number of layers and number of neurons are decided in the different layers and how these layers are interconnected with each other through corresponding weights. The numbers of inputs and output layers are always constant, that is, a single layer is required at the input and output sides but it is very tough task to determine the number of hidden layers and the number of neurons in it. The rule-of-thumb method is used for deciding the number of neurons in the hidden nodes. According to this rule, the total numbers of hidden layer neurons are summation of the size of output layer plus 2/3 of the input layer size. The number of hidden layer neurons should be less than twice of the number of neurons in input layer. The size of the hidden layer neurons is between the input layer size and the output layer size. But the above three methods for determining the number of hidden nodes are not always true because not only the input layer and the output layer determines the size of the hidden layer neurons but also the dataset of training samples, types of architecture, the training algorithm, and complexity of the activation function applied on the neurons decides it. Multiple hidden layers are used in the applications where accuracy is the criteria and training time is not important. Usually, one or two hidden layers are adequate for resolving any nonlinear complex task. The third hidden layer can be added for improving accuracy, but this will increase overall complexity of the neural network and the total training time will be increased. Four hidden layers are not used in the neural network architecture because it cannot follow the rule of thumb. If there are three hidden layers then total neurons in hidden layers $= 2\ (3 \times 2/3)$ (twice of input). Minimum hidden layer must be used in ANN to reduce the complexity and training time. As shown in **Figure 10**, neural network architecture consists of

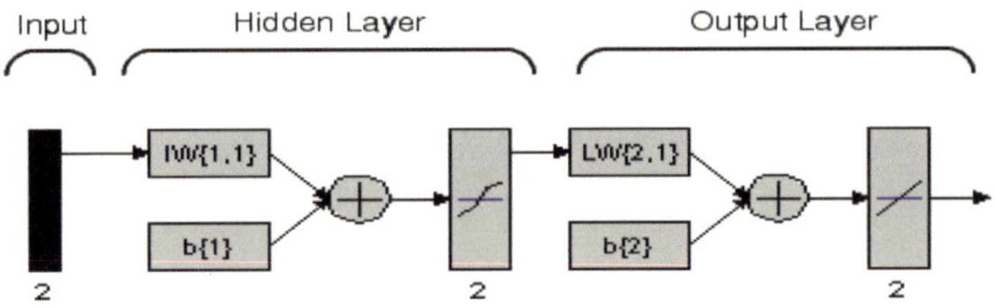

Figure 10. Neural network architecture for microstrip patch antenna.

three layers. It has one input, one hidden, and one output layer. Input, hidden, and output layer consists of two neurons. Inputs W and L are applied at the input neurons while outputs f_1 and f_2 are obtained from the output neurons. The ANN training graph results for the analysis of microstrip patch antenna are shown in **Figure 11**, which shows that 21 epochs are required for ANN training. The full sets of input samples are passed through the artificial neural network for minimization of the error in the back-propagation algorithms, which is known as an *epoch*. Training graph reveals that error reduced from 10^1 to nearly 10^{-4}, as shown in **Figure 11**.

Trained results (f_1 and f_2) for ANN analysis of patch are given in **Table 2**.

2.1.5. Synthesis of microstrip patch antenna using artificial neural network

Synthesis is defined as to obtain the dimensions of microstrip patch (W, L) as targets while providing the cutoff frequencies (f_1, f_2), thickness and dielectric constants of the dielectric material as the input parameters of ANN model. Around 25–30 samples are collected for ANN training which is obtained by varying the dimensions of microstrip patch antenna using electromagnetic simulation software as illustrated in **Table 3**. It is used for the synthesis of patch antenna. Trained Results (W, L) for ANN Synthesis of patch is illustrated in **Table 4**. The neural network is trained by using back-propagation algorithm. The ANN training graph

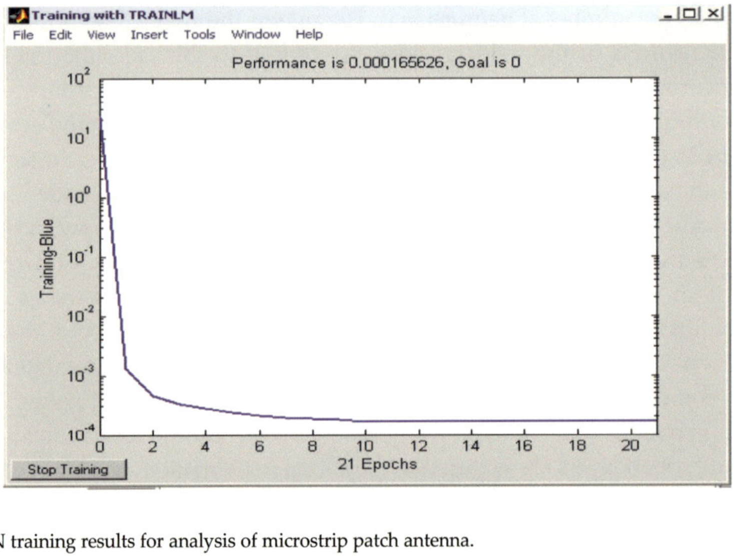

Figure 11. ANN training results for analysis of microstrip patch antenna.

Inputs (dimensions of patch in mm)		Targets (lower and upper cutoff frequency in GHz)	
W	*L*	f_1	f_2
17.7	13.34	4.9308	5.038
17.7	13.55	4.8705	4.9605
17.7	13.65	4.8512	4.9402
17.7	13.85	4.808	4.8948
17.7	14.05	4.7594	4.8438
17.7	14.15	4.7335	4.8165
17.7	14.25	4.7067	4.7884
17.7	14.35	4.6793	4.7596
17.7	14.45	4.6517	4.7305
18.3	13.85	4.7806	4.866
18.3	14.35	4.6474	4.726
18.8	14.35	4.6154	4.6924
18.8	14.85	4.4872	4.5577
19.3	14.85	4.4617	4.5309
19.3	15.35	4.3742	4.439
19.8	15.35	4.3592	4.4232
20.3	15.85	4.3042	4.3653
20.3	16.35	4.279	4.3385
20.8	16.35	4.2735	4.3324
20.8	16.85	4.2466	4.3027
21.3	16.85	4.2351	4.2897
21.3	17.35	4.1668	4.2123
21.8	17.35	4.1367	4.1781
21.8	17.85	4.0197	4.045
22.3	17.85	3.9887	4.0098
22.3	18.35	3.919	3.930

Table 2. Trained results for ANN analysis of patch.

Inputs (lower and upper cutoff frequency in GHz)		Targets (dimensions of patch in mm)	
f_1	f_2	*W*	*L*
4.94	5.04	17.7	13.34
4.86	4.95	17.7	13.55
4.82	4.91	17.7	13.65
4.8	4.89	17.7	13.85
4.77	4.85	17.7	14.05

Inputs (lower and upper cutoff frequency in GHz)		Targets (dimensions of patch in mm)	
f_1	f_2	W	L
4.73	4.81	17.7	14.15
4.71	4.79	17.7	14.25
4.69	4.76	17.7	14.35
4.66	4.73	17.7	14.45
4.78	4.87	18.3	13.85
4.65	4.73	18.3	14.35
4.61	4.7	18.8	14.35
4.49	4.55	18.8	14.85
4.47	4.55	19.3	14.85
4.37	4.41	19.3	15.35
4.35	4.41	19.8	15.35
4.31	4.37	20.3	15.85
4.29	4.35	20.3	16.35
4.27	4.33	20.8	16.35
4.26	4.33	20.8	16.85
4.21	4.27	21.3	16.85
4.16	4.21	21.3	17.35
4.14	4.19	21.8	17.35
4.02	4.04	21.8	17.85
3.99	4.01	22.3	17.85
3.92	3.93	22.3	18.35

Table 3. Simulated results for ANN synthesis of patch.

Inputs (lower and upper cutoff frequency in GHz)		Targets (dimensions of patch in mm)	
f_1	f_2	W	L
4.94	5.04	17.6823	13.2641
4.86	4.95	17.6905	13.751
4.82	4.91	17.7813	13.8311
4.8	4.89	17.8317	13.8756
4.77	4.85	17.8594	13.9008
4.73	4.81	17.9753	14.0034
4.71	4.79	18.0396	14.0602
4.69	4.76	18.0393	14.0613
4.66	4.73	18.1438	14.154

Inputs (lower and upper cutoff frequency in GHz)		Targets (dimensions of patch in mm)	
f_1	f_2	*W*	*L*
4.78	4.87	17.8858	13.9234
4.65	4.73	18.2604	14.2558
4.61	4.7	18.5302	14.4931
4.49	4.55	18.8415	14.7789
4.47	4.55	19.2449	15.1307
4.37	4.41	19.3158	15.2211
4.35	4.41	19.8384	15.6756
4.31	4.37	20.1946	15.9979
4.29	4.35	20.3849	16.1706
4.27	4.33	20.5834	16.3509
4.26	4.33	20.9421	16.6629
4.21	4.27	21.2251	16.9371
4.16	4.21	21.4873	17.2026
4.14	4.19	21.7157	17.4157
4.02	4.04	21.885	17.7323
3.99	4.01	22.1746	18.0291
3.92	3.93	22.3253	18.3111

Table 4. Trained results for ANN synthesis of patch.

results for the synthesis of microstrip patch antenna is shown in **Figure 12** which shows that 100 epochs are required for ANN training and error get reduced from 10^2 to nearly 10^{-2}.

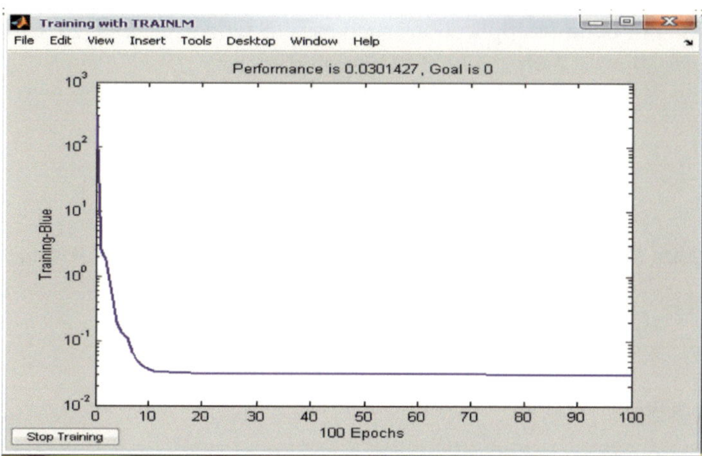

Figure 12. ANN training results for synthesis of microstrip patch antenna.

Network errors and weights are obtained in matrix form during ANN training as given below.

Network errors:

[0.20766 0.10947 0.018694 −0.031744 −0.05942 −0.17535 −0.23957 −0.2393 −0.34378 0.41415 0.039568 0.26975 −0.041452 0.055095 −0.015806 −0.038441 0.1054 −0.084949 0.21663 −0.14213 0.074867 −0.18731 0.084323 −0.085043 0.12542 −0.025298;

−0.31411 −0.20102 −0.1811 −0.025609 0.14918 0.14663 0.1898 0.28873 0.296 −0.073367 0.094171 −0.14313 0.071068 −0.28069 0.12893 −0.32561 −0.14788 0.17943 −0.00094985 0.18709 −0.087086 0.14741 −0.065711 0.11766 −0.17911 0.038876]

Weights

[1.0349 3.4722;

3.0377 2.2758]

3. Conclusion

In this work, the rectangular microstrip patch antenna is designed using the artificial neural network modeling procedure. Here synthesis refers to forward side and analysis refers to reverse side of the problem. Therefore in synthesis problem, the geometric dimensions such as length and the width of antenna are obtained with more accuracy in less time as compared to simulation software while providing resonant frequency, thickness, and dielectric constant at the input side of the ANN model. In the analysis problem of patch antenna, resonant frequency or both upper and lower cutoff frequencies of patch antenna are obtained at the output side of the ANN model while providing the dimensions of patch (W, L) and other parameters at the input side of the ANN model with much accuracy in less time.

Now the future work in this work includes trying more topologies to obtain more compact patch antennas, filters, and many other microwave/RF modeling design using artificial neural network for different band of applications such as ultra-wideband (UWB), Global System for Mobile communications (GSM), and WiMAX applications.

Appendix

MATLAB program for finding out the dimensions of microstrip patch antenna using the ANN

```
NNTWARN OFF

disp_freq = 100;

max_epochs = 8000;

%1r = 0.013;

err_goal = 0.009;
```

p = [4.94 4.86 4.82 4.8 4.77 4.73 4.71 4.69 4.66 4.78 4.65 4.61 4.49 4.47 4.37 4.35 4.31 4.29 4.27 4.26 4.21 4.16 4.14 4.02 3.99 3.92;

5.04 4.95 4.91 4.89 4.85 4.81 4.79 4.76 4.73 4.87 4.73 4.7 4.55 4.55 4.41 4.41 4.37 4.35 4.33 4.33 4.27 4.21 4.19 4.04 4.01 3.93];

tx = [17.7 17.7 17.7 17.7 17.7 17.7 17.7 17.7 17.7 18.3 18.3 18.8 18.8 19.3 19.3 19.8 20.3 20.3 20.8 20.8 21.3 21.3 21.8 21.8 22.3 22.3;

13.35 13.55 13.65 13.85 14.05 14.15 14.25 14.35 14.45 13.85 14.35 14.35 14.85 14.85 15.35 15.35 15.85 16.35 16.35 16.85 16.85 17.35 17.35 17.85 17.85 18.35];

%bus6

[w1,b1,w2,b2] = initff(p,2,'tansig',tx,'purelin');

Tp = [disp_freq,max_epochs,err_goal];

[w1,b1,w2,b2,epochs,tr] = trainbpx(w1,b1,'tansig',w2,b2,'purelin',p,tx,tp);

gil = simuff(p,w1,b1,'tansig',w2,b2,'purelin')

save patch1.mat;

Find the output:

P = [4.94 ; 5.04];

Result = simuff(p,w1,b1,'tansig',w2,b2,'purelin')

outputs:

[17.683 ; 13.265]

Author details

Vivek Singh Kushwah[1]* and Geetam Singh Tomar[2]

*Address all correspondence to: vivek_kushwah@rediffmail.com

1 Amity University, Madhya Pradesh, India

2 Govt. THDC Engg. College, Tihri, Uttarakhand, India

References

[1] Balanis CA. Antenna Theory. United States: John Wiley & Sons, Inc; 1997

[2] Bahl IJ, Bhartia P. Microstrip Antennas. Dedham, MA: Artech House; 1980

[3] Zhang QJ, Gupta KC. Neural Networks for RF and Microwave Design. USA: Artech House Publishers; 2000

[4] Wood C. Improved bandwidth of microstrip antennas using parasitic elements. IEEE Proceedings on Microwaves, Optics and Antennas. 1980;**127**(4):231–234

[5] Dey S, Mittra R. Compact microstrip patch antenna. Microwave and Optical Technology Letters. 1996;**13**(1):12–14

[6] Wong KL, Tang CL, Chen HT. A compact meandered circular microstrip antenna with a shorting pin. Microwave and Optical Technology Letters. 1997;**15**(3):147–149

[7] Sağıroğlu S, Güney K, Erler M. Resonant frequency calculation for circular microstrip antennas using artificial neural networks. International Journal of RF and Microwave Computer-Aided Engineering. 1998;**8**(3):270–277

[8] Karaboğa D, Güney K, Sağıroğlu S, Erler M. Neural computation of resonant frequency of electrically thin and thick rectangular microstrip antennas. Microwaves, Antennas and Propagation, IEEE Proceedings. 1999;**146**(2):155–159

[9] Kuo YL, Chiou TW, Wong LK A novel dual-band printed inverted-F antenna. Microwave and Optical Technology Letters. 2001;**31**(5):353–355

[10] Reed S, Desclos L, Terret C, Toutain S. Patch antenna size reduction by means of inductive slots. Microwave and Optical Technology Letters. 2001;**29**(2):79–81

[11] Anguera J, Puente C, Borja C, Montero R, Soler J. Small and high-directivity bow-tie patch antenna based on the Sierpinski fractal. Microwave and Optical Technology Letters. 2001;**31**(3):239–241

[12] Güney K, Sağıroğlu S, Erler M. Generalized neural method to determine resonant frequencies of various microstrip antennas. International Journal of RF and Microwave Computer-Aided Engineering. 2002;**12**(1):131–139

[13] Mishra RK, Patnaik A. Designing rectangular patch antenna using the neurospectral method. Antennas and Propagation, IEEE Transactions. 2003;**51**(8):1914–1921

[14] Tqürker N, Günes F, Yildirim T. Artificial neural design of microstrip antennas. Turkish Journal of Electrical Engineering & Computer Sciences. 2006;**14**(3):445–453

[15] Ansari JA, Yadav NP, Mishra A, Kamakshi, Singh A. Broadband rectangular microstrip antenna loaded with a pair of u-shaped slot. International Journal of Microwave and Optical Technology. 2011;**6**(4):185–190

[16] Garima, Bhatnagar D, Saini JS, Saxena VK, Joshi LM. Design of broadband circular patch microstrip antenna with diamond shapeslot. Indian Journal of Radio & Space Physics. 2011;**40**:275–281

[17] Wang Z, Fang S, Fu S, Jia S. Single-fed broadband circularly polarized stacked patch antenna with horizontally meandered strip for universal UHF RFID applications. IEEE Transactions on Microwave Theory and Techniques. 2011;**59**(4):1066–1073

[18] Yadav D. L-slotted rectangular microstrip patch antenna. In: Proceedings of the IEEE International Conference on Communication Systems and Network Technologies; 3–5 June 2011; Jammu, India. US: IEEE; 2011. pp. 220–223

[19] Ali MT, Aizat, Pasya I, Mazlan Zaharuddin MH, Yaacob N. E-shape microstrip patch antenna for wideband applications. In: Proceedings of the IEEE International RF and Microwave Conference; 12–14 December 2011; Negeri Sembilan, Malaysia. US: IEEE; 2011. pp. 439–443

[20] Varma R, Ghosh J. Design of slot loaded proximity coupled microstrip antennas using knowledge based neural networks. In: Proceedings of the IEEE International Conference on Recent Advances in Information Technology; 3–5 March 2016; Dhanbad, India. US: IEEE; 2016. pp. 209–214

Section 3

Recent Trends to Microstrip Antennas Research

Chapter 4

Circularly Polarized T-Stub Coupled Microstrip Antenna Structure for WLAN

Milind Thomas Themalil

Additional information is available at the end of the chapter

http://dx.doi.org/10.5772/67931

Abstract

In this chapter, a novel feeding mechanism for single layer microstrip patch antenna to generate circular polarization using electromagnetically coupled microstrip T junction is presented. The antenna structure eliminates the need for capacitors in the RF path for active antenna applications in wireless local area network (WLAN). The simulated results were verified by measurement using the vector network analyzer.

Keywords: active antenna, DC isolation, T-stub coupled, circularly polarized

1. Introduction

Wireless local area networks (WLAN) use antennas with different standards having characteristics such as miniaturization, conformability, wideband operation, multiple resonances and good gain. New microstrip antenna designs hence provide an ideal solution due to these features [1, 2]. Active antennas integrate the circuit with radiation properties into a monolithic microwave integrated circuit; a significant research area in recent years. Broadband wireless communications have increased rapidly in recent years demanding quality of service, security, handover and increased throughput for the wireless local area networks. The aim of next generation wireless communication is high-speed networking service enabling multimedia communication. An important high data rate wireless broadband communication standard is ETSI high performance local area network type 2 (HIPERLAN 2) which uses 5.15–5.725 GHz band with omni-directionality. Dual circularly polarized antennas, as far as active integrated antennas are concerned, are compact and single layered.

A drawback of single layered patch antenna structures, whether probe fed or microstrip line fed, for active antennas with dual polarization, is their DC contact between ports via the patch.

Active antennas require DC isolated ports. A solution is to embed additional capacitors in the RF path or use multilayered structures like aperture coupling. High Q lumped RF capacitors are monetarily expensive and complicated to model in the design process of active antennas, with an increase in the physical size and cross polarization levels. Circular polarization provides flexibility in antenna orientation between transmitter and receiver, better mobility, weather penetration features and reduction in multipath fading effects. However, some inherent limitations include the attainable impedance and axial ratio bandwidths [3, 4]. Plenty of investigations have been carried out on this subject, which include design of circularly polarized microstrip antennas with single feed and narrow axial ratio bandwidth and also with double feed with relatively wide axial ratio bandwidth. The structure presented in this chapter is capable of providing DC as well as RF isolation, direct 50 Ω matching and increased bandwidth compared to other single layer feeding mechanisms. The antenna is circularly polarized in broadside direction which is an attractive feature for WLAN applications. Good gain and axial ratio bandwidth are attained.

2. Design of the antenna structure

The antenna structure presented in this chapter consists of two-coupled microstrip T junctions on adjacent sides of a square patch as shown in **Figure 1**. The T junctions transform the high radiation resistance of the patch to lower impedance [5]. The microstrip Ts have DC isolation between them; there is no need for capacitors in the RF path in dual polarized active antenna applications [6, 7]. The small width of the microstrip T causes the currents in the two

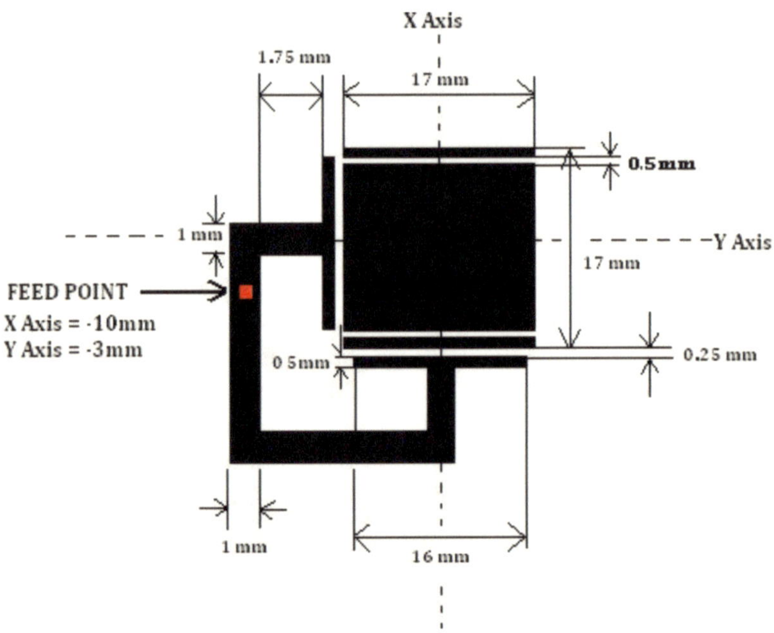

Figure 1. Top view of the microstrip T-stub coupled patch antenna.

arms of the microstrip Ts to be in opposite directions to the excited TM modes. Therefore, the radiation pattern of the structure is not affected significantly by the microstrip T. By controlling the length of the T arm and the distance between the patch and the T arm, we can achieve 50 Ω impedance match. Dimension of the patch is 16.8 mm × 16.8 mm. The T-arms provided on both radiating and non-radiating edge are fed by a co-axial probe. Two rectangular slots loaded on the patch on opposite sides provide circular polarization as in **Figure 1**. Slot loading on opposite edges significantly improves bandwidth besides providing circular polarization [8]. The co-axial probe feed point is 10 mm to the left of centre and 3 mm below the centre of the patch. In this simulation, a 6 dBi gain bandwidth of 350 MHz is obtained.

The antenna was fabricated on a 1.6 mm thick PTFE substrate with dielectric constant $\varepsilon_r = 2.32$ and loss tangent tanδ = 0.001. The fabricated antenna prototype is shown in **Figure 2**.

Figure 2. Photograph of fabricated microstrip T-stub coupled patch antenna.

3. Simulated and measured results for the structure

The simulated and measured resonance frequencies are 5.22 and 5.205 GHz as shown in **Figures 3** and **4**, respectively. The simulated and measured return losses are −27 and −25 dB as shown in **Figures 3** and **4**, respectively. Simulated axial ratio is shown in **Figure 5**. Minimum axial ratio is 0.2 dB. The simulated and measured −10 dB impedance bandwidths are 380 MHz as shown in **Figure 6** and 350 MHz as shown in **Figure 4**.

The simulated results of radiation pattern and gain of the structure are presented in **Figures 7** and **8**, respectively. Radiation is in the broadside direction and the highest gain obtained is 6.5 dBi. The 6 dBi gain bandwidth product is 350 MHz which is well within the range of the WLAN application in the 5.2 GHz band. A good gain is obtained throughout the entire range of application and the antenna has 77% antenna efficiency as shown in **Figure 9**.

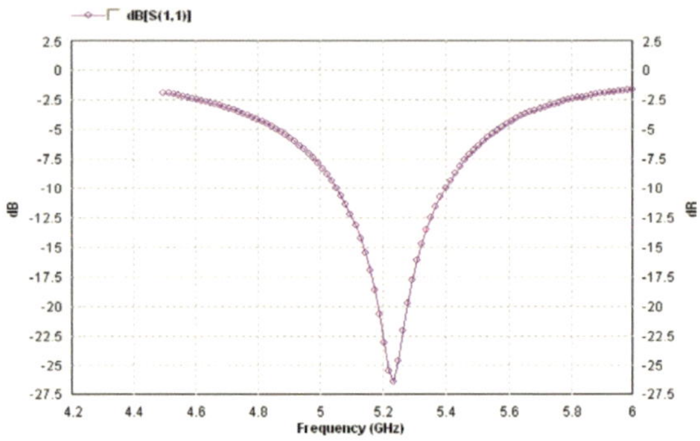

Figure 3. Simulated return loss.

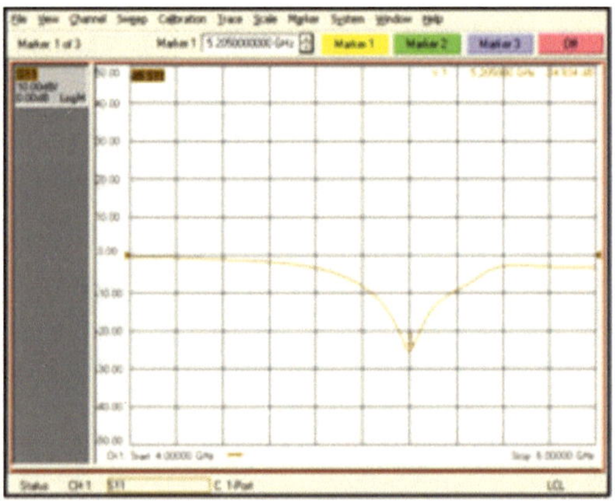

Figure 4. Measured return loss.

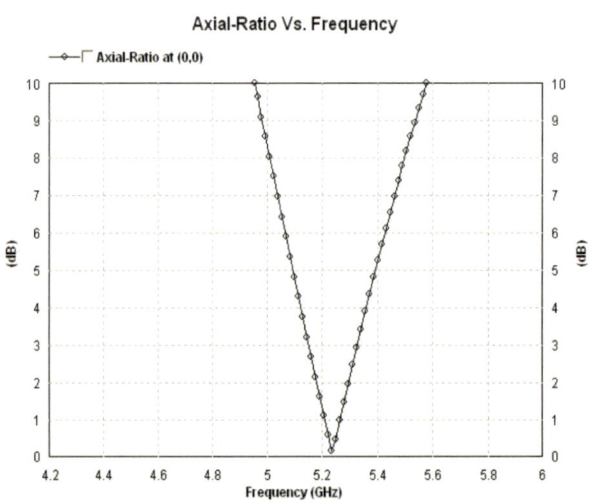

Figure 5. Simulated axial ratio.

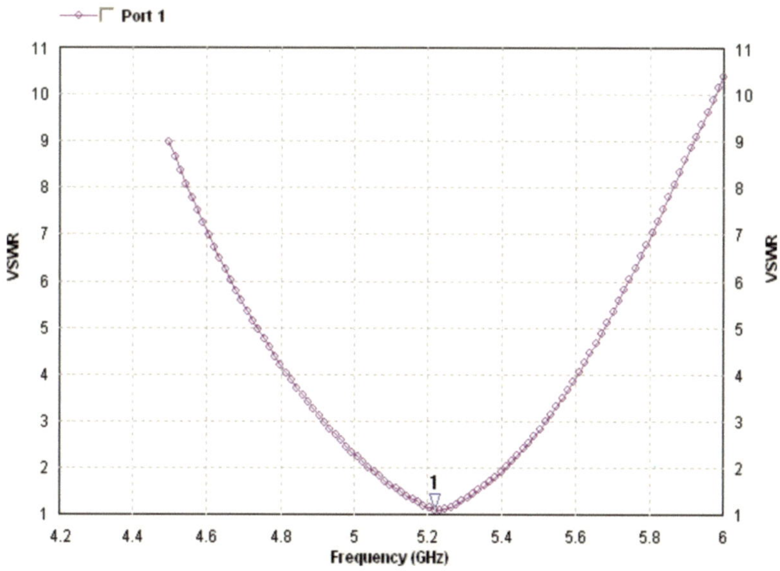

Figure 6. Simulated VSWR.

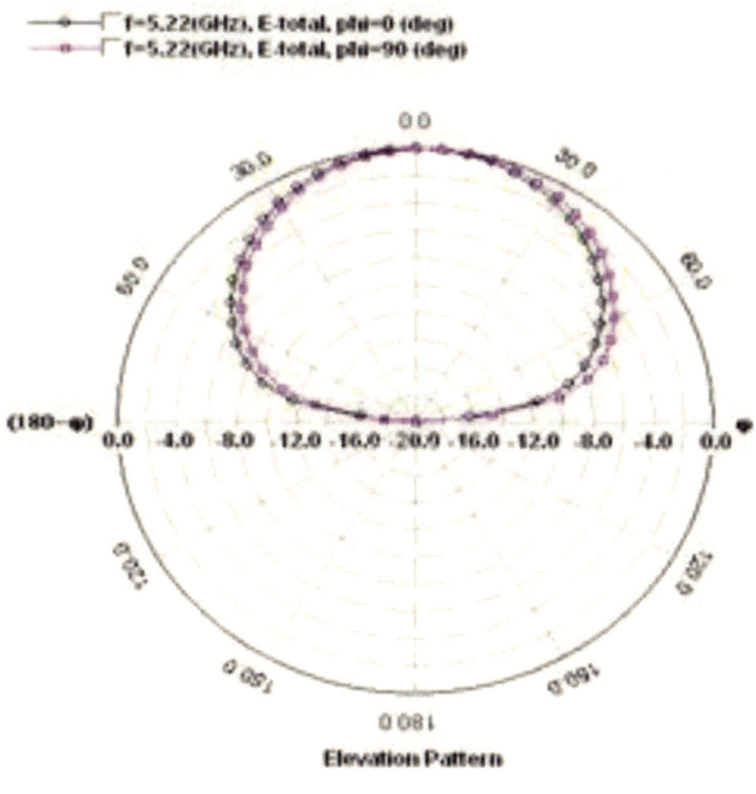

Figure 7. Simulated radiation pattern.

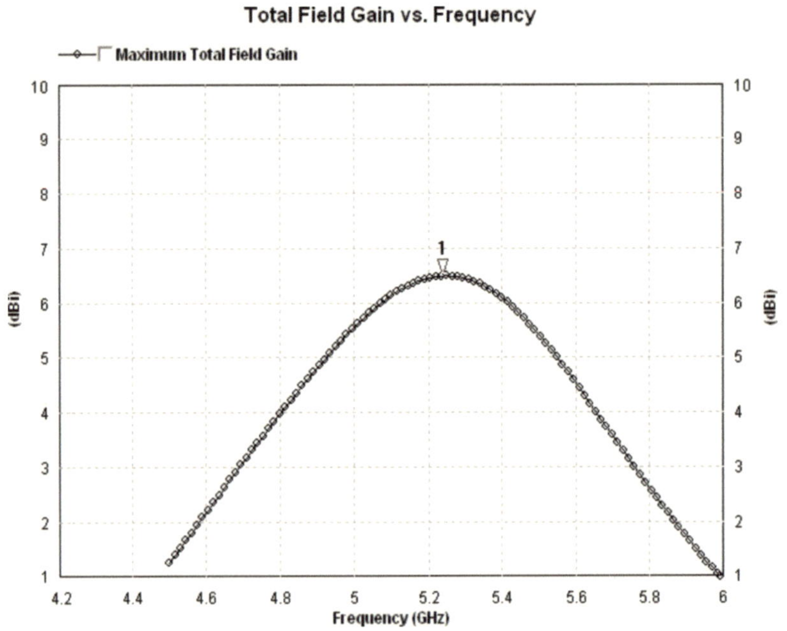

Figure 8. Simulated gain.

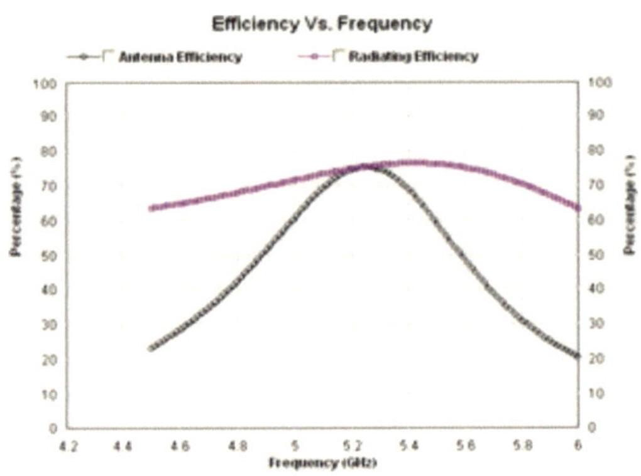

Figure 9. Simulated efficiency.

Tables 1 and **2** show the simulated and measured results.

Resonance frequency	Return loss S_{11}	10 dB Return loss bandwidth	Max. gain (dBi)	6 dBi gain bandwidth	Min. axial ratio	3 dB axial ratio bandwidth	Efficiency
5.22 GHz	−27 dB	380 MHz	6.5 dBi	350 MHz	0.2 dB	180 MHz	77%

Table 1. Simulated results.

http://dx.doi.org/10.5772/67931

Resonance frequency	Return loss S_{11}	10 dB return loss bandwidth
5.205 GHz	−25 dB	350 MHz

Table 2. Measured results.

4. Conclusion

A compact and broadband circularly polarized microstrip patch antenna structure with T-shaped electromagnetically coupled ports was designed, simulated, fabricated and tested. The antenna structure is intended for active antenna WLAN applications. The simulated results are verified by the measurement using the vector network analyzer. For active antenna applications, the structure does not need any capacitor in the RF path of the circuit.

Acknowledgements

The author would like to acknowledge the work of Dr. Anders Rydberg and Dr. Dhanesh Kurup of the Signals & Systems Division, Uppsala University, Sweden for inspiring the present chapter. Deepest gratitude to Dr. J S Roy of KIIT University, Bhubaneswar, India and Dr. Bhaskar Gupta of Jadavpur University, Kolkata, India for supervising the author's Ph.D Thesis. The author would like to thank the Electronics & Communication Engineering Department of Birla Institute of Technology, Ranchi, India and the Electronics & Telecommunication Engineering Department of Jadavpur University, Kolkata, India for their research facilities. The author would also like to acknowledge the Microwave Laboratory Division, Department of Electronics and Communication Engineering, Indian Institute of Science (IISc.), Bangalore, India for their research and measurement facilities. The author is extremely grateful to Believers Church Caarmel Engineering College, Kerala, India for their inspiration, encouragement and support.

Author details

Milind Thomas Themalil

Address all correspondence to: milindthomas@gmail.com

Research & Development, Believers Church Caarmel Engineering College, Kerala, India

References

[1] James JR, Hall PS. Handbook of Microstrip Antennas, London: Peter Peregrinus, 1989

[2] Pozar DM, Schaubert DH. Microstrip Antennas: Analysis and Design of Microstrip Antennas and Arrays, New York: IEEE Press, 1995

[3] Chung KL. A wideband circularly polarized H-shaped patch antenna, IEEE Transactions of Antennas and Propagation. 2010;**58**:3379-3383

[4] Chang FS, Wong KL, Chiou TW. Low-cost broadband circularly polarized patch antenna. IEEE Transactions of Antennas and Propagation. 2003;**51**:3006-3009

[5] Kurup DG, Rydberg A, Himdi M. Compact microstrip T – coupled patch antenna for dual polarization and active antenna applications. Electronics Letters. 2002;**38**(21):1240-1241

[6] Choi DH, Park SO. Active integrated antenna using a T shaped microstrip coupled patch antenna. Microwave & Optical Technology Letters. 2005;**44**(5):434-436

[7] Chang K, Hummer KA, Gopalakrishnan GK. Active radiating element using FET source integrated with microstrip patch antenna. Electronics Letters. 1988;**24**:1347-1348

[8] Roy JS, Thomas M. Design of a circularly polarized microstrip antenna for WLAN. Progress In Electromagnetic Research, PIER M. 2008;**3**:79-90

Chapter 5

Bio-Inspired Microstrip Antenna

Alexandre Jean René Serres,
Georgina Karla de Freitas Serres,
Paulo Fernandes da Silva Júnior,
Raimundo Carlos Silvério Freire,
Josiel do Nascimento Cruz,
Tulio Chaves de Albuquerque,
Maciel Alves Oliveira and
Paulo Henrique da Fonseca Silva

Additional information is available at the end of the chapter

http://dx.doi.org/10.5772/intechopen.69766

Abstract

In the last few years, bio-inspired solutions have attracted the attention of the scientific community. Several world-renowned institutions have sponsored and created laboratories in order to understand the forms, functions and behavior of living organisms. Some methods can be highlighted in the search for geometric representation of the shapes found in the nature, the fractal geometry, the polar geometry, and the superformula of Gielis. This chapter is focusing on bio-inspired microstrip antennas, especially on leaf-shaped antennas from the Gielis superformula that open a vast research field for more compact antennas with low visual impact.

Keywords: bio-inspired, Gielis superformula, compact antenna, leaf-shaped antennas, visual impact

1. Introduction

From the 1990s, with the advent of the Internet, the popularization of portable terminals (laptops, mobile phones, etc.) favored the telecommunications industry and the infrastructure of networks experienced a remarkable growth [1, 2]. When the information age emerges from an increasingly networked world, the digital information, and communication technology permeate the society and are increasingly important to their development [3, 4]. Modern wireless applications demand esthetic, multifunctional, and portable terminals that operating in multiple frequency bands and can integrate different wireless services: 4G, Wi-Fi, Bluetooth, NFC, GPS and so on. Future trends toward 5G systems also require enhanced mobile broadband for emergent applications [5].

With the rapid advance of wireless communication systems, the use of antennas in base stations and portable terminals must meet increasingly stringent criteria, such as miniaturization, integration with other systems, and multiband or broadband operation [1–4]. Due its attractive features, low-profile microstrip antennas (MSA) and arrays are well suitable to meet the demands of fixed or mobile wireless applications [6–9].

Antenna parameter specifications change according to application. Indeed, fixed antennas must have high gain, stable radiation pattern, and bandwidth tolerance; embedded antennas should be efficient in radiation and possess larger beam width [3]. In short-range UWB wireless systems, the antenna bandwidth exceeds the lesser of 500 MHz or 20% of the center frequency [9]. Thus, impedance bandwidth, gain, radiation pattern, and polarization are fundamental parameters for antenna designers to take into account.

A trend in the application of antennas for modern wireless systems is the use of compact antennas with stable radiation coverage over a wideband [2–4]. An antenna must be compact in many situations: embedded antennas, wearable antennas, camouflaged antennas, and so on. However, most often an antenna electrically small narrows the impedance bandwidth, reduces gain, and limits control of the resulting radiation pattern [4, 6, 9].

Various institutions around the world have invested resources and established research centers with focus on biologically inspired engineering as the Massachusetts Institute Technology, the London College, and the Harvard University, for example. This research branch looks in nature similar solution to the problems encountered in engineering. With appropriate adjustments, it is possible to adapt the solutions used by the nature of the engineering problems [10]. According to [11], the development of a bioinspired methodology requires three steps: identification of analogies, with structures and methods that are similar; understanding, detailed modeling of actual biological behavior; and engineering, which is the process of model simplification and adjustments to technical applications.

The researches that use the bio-inspired geometry for the development of antennas are recent and can be divided into two groups: antennas with bio-inspired geometries in animals and antennas with bio-inspired geometries in plants. Bio-inspired antennas in animals try to use internal organs or external parts of the animals which work analogously to the operation of the antennas used in the communication systems. A biomimetic antenna in the shape of a bat's ear [12] can be cited, a biologically inspired electrically small antenna arrays that mimic the hearing mechanism of such insects [13], an antenna system based on the wasp's curved antennas [14] and a biologically inspired vascular antenna reconfiguration mechanism [15]. Research on antennas with bio-inspired plant geometries uses the plants or part of them (stem, leaves, and flowers) to develop antennas for various frequencies and technologies. The study of microstrip antennas with models bio-inspired on leaves (leaf-shaped antennas) has aroused the interest of researchers due to the good results. The leaves present similar characteristics to fractals as, for example, the reduction of the total dimensions with the increase of perimeter. Furthermore, the leaves have a light-harvesting reaction center complex, i.e., an array of antennas capable of operating in the visible light range (400–700 nm) with characteristics which analog to satellite dishes. The main purpose of the leaf shape is to capture the sunlight to transform it in chemical energy by a photosynthesis process. The bioinspired on leaves open a vast research field for more compact antennas with low visual impact. Among the published works, the following can be highlighted. A leaf-shaped monopole antenna with an extremely wide bandwidth is introduced in Ref. [16] and a leaf-shaped bowtie slot antenna for UWB applications in Ref. [17]. A band-notched tulip antenna for UWB applications is presented in Ref. [18] and a wide-band tulip-loop antenna in Ref. [19]. More recently, a MIMO antenna using castor leaf-shaped quasi-self-complementary elements for broadband applications and a bio-inspired design of directional leaf-shaped printed monopole antennas for 4G 700 MHz band are presented in Refs. [20] and [21], respectively.

This chapter discusses the design of innovative bio-inspired microstrip antennas with fractal, polar, and Gielis shapes, which are optimized for wireless applications. Section 2 presents the different methods to generate bio-inspired shapes. In section 3 are presented some applications including esthetic wearable antennas and antenna arrays. In this last section, the results obtained from the Gielis superformula are highlighted due to the simplicity and flexibility of the formulation. Simulations are performed using ANSYS Electronics Desktop. Measurements of prototypes are compared to simulations and classical designs as circular printed monopole antenna in some cases.

2. Bio-inspired shapes

2.1. Fractal and polar transformations

From a mathematical point of view, a fractal refers to a set in Euclidean space with specific properties, such as self-similarity or self-affinity, simple and recursive definition, fractal dimension, irregular shape, and natural appearance [22]. Fractal geometry is the study of sets with these properties, which are too irregular to be described by calculus or traditional Euclidian geometry language [22, 23].

Fractals are resort to conventional classes, such as geometrical fractals, algebraic fractals, and stochastic fractals [24]. Two common methods used to generate mathematical fractals are iterated function systems (IFS) and Lindenmayer systems [22–25].

Lindenmayer system (or L-system) was initially conceived to model growth phenomena in biological organisms [26]. An L-system grammar performs an initial string of symbols (axiom) and includes a set of production rules that may be applied to the symbols (letters of the L-system alphabet) to generate new strings.

In **Figure 1** are shown four examples of fractal iterations using IFS and L-system.

Like fractals, polar transformations give rise to a wide class of shapes. A polar transformation is defined in this chapter through a vector function $\vec{v}(t) = (x(t), y(t)),\ t \geq 0$, that is, for each real value, t is associated with a vector in $\Re^2$, Eq. (1). An example of an esthetic polar transformation is defined by Eq. (2) and presented in **Figure 2** for k varying up to $k = 24$ petals.

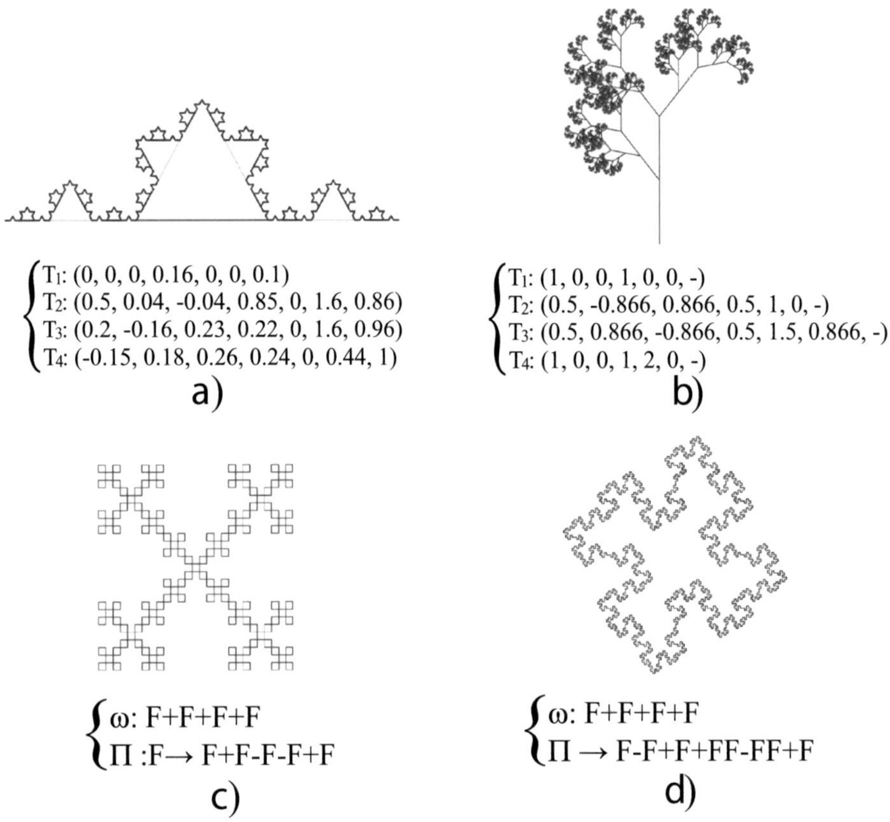

Figure 1. IFS and L-system prefractals: (a) Koch curve; (b) modified Barnsley fern; (c) Koch island; (d) Minkowiski island.

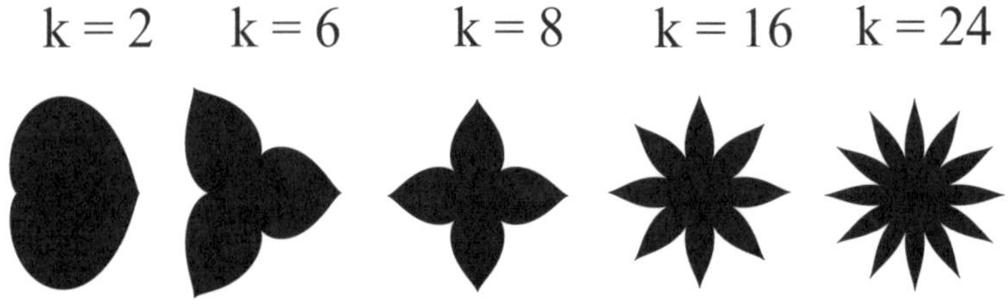

Figure 2. Esthetic space-filling polar transformation for k varying up to $k = 24$.

http://dx.doi.org/10.5772/intechopen.69766

$$\begin{aligned} \vec{v}(t) &: I \rightarrow \mathfrak{R}^n \\ t &\rightarrow \vec{v}(t) \end{aligned} \quad (1)$$

$$\vec{v}(t) = \left(1 + \frac{\cos(t)}{2}\right) \cdot \left(\cos\left(\frac{2t - sen(2t)}{k}\right), \sin\left(\frac{2t - sen(2t)}{k}\right)\right), \quad 0 \le t \le k\pi \quad (2)$$

2.2. Gielis superformula

The superformula is a generalization of the superellipse and was first proposed by Johan Gielis in 2003 [27]. With this model, it is possible to describe mathematically a wide variety of natural and abstract forms, such as leaf and flower shapes, for example. Gielis started from the concept of superellipses (3) and obtained the superformula (4), which is based on the idea that many natural forms can be interpreted as modified circles. To obtain this result, it was used polar coordinates, replacing $x = r \,.\, \cos(\varphi)$ and $y = r \,.\, \sin\varphi$ in addition to introducing the argument $m/4$, which confers rotational symmetry in some structures, and the possibility of using different values of exponent n for each term (through n_1, n_2, n_3).

$$\left|\frac{x}{a}\right|^n + \left|\frac{y}{b}\right|^n = 1 \quad (3)$$

$$r(\varphi) = \frac{1}{\left\{\left[\left(\left|\frac{1}{a}\cos\left(\varphi\frac{m}{4}\right)\right|\right)^{n2} + \left(\left|\frac{1}{b}\sin\left(\varphi\frac{m}{4}\right)\right|\right)^{n3}\right]^{\frac{1}{n1}}\right\}} \quad (4)$$

From the manipulation of the six parameters (a, b, m, n_1, n_2, n_3) of (4), called the Gielis superformula, it is possible to generate and modify several shapes. The superformula can also be multiplied by other mathematical functions, generating other forms. In order to illustrate the possibilities of the superformula, some examples of star shaped, leaf shaped, butterfly shaped, and flower shaped were generated. These shapes can be seen, with all the parameters used, in **Tables 1–4**, respectively.

	Superformula parameters							
Shape	a	b	m	n_1	n_2	n_3	Function	ϕ
	1	1	7	10	6	6	1	$0{:}2\pi$
	10	10	5	2	7	7	1	$0{:}2\pi$
	1	1	5	2	13	13	1	$0{:}2\pi$

Table 1. Star-shaped geometries.

	Superformula parameters							
Shape	*a*	*b*	*m*	n_1	n_2	n_3	**Function**	ϕ
	1	1	2	500	1000	1000	1	0:2π
	2	2	3	1	1	1	1	0:2π
	2	2	2	1	1	2	1	0:2π

Table 2. Leaf-shaped geometries.

	Superformula parameters							
Shape	*a*	*b*	*m*	n_1	n_2	n_3	**Function**	ϕ
	1	1	4	0.2	0.5	15	$\lvert\cos(m\varphi)\rvert$	0:2π
	1	1	4	0.2	0.5	15	1	0:2π
	1	1	4	0.2	0.5	15	$\lvert\cos(2m\varphi)\rvert$	0:2π

Table 3. Butterfly-shaped geometries.

	Superformula parameters							
Shape	*a*	*b*	*m*	n_1	n_2	n_3	**Function**	ϕ
	1	1	10	0.1	10	0.01	1	0:2π
	10	1	100	1	0	4	1	0:2π
	1	1	2.5	100	2.7	2.7	$\lvert\cos(m\varphi)\rvert$	0:2π

Table 4. Flower-shaped geometries.

3. Applications in microstrip antennas

The bio-inspired shapes are generated using the software MATLAB© in format DXF.

The simulations were performed with the commercial software ANSYS Electronics Desktop™, and the measurements in the Radiometry Laboratory of the Federal University of Campina Grande, with the VNA Agilent Technologies, model E5071C-280 (9 kHz–8.5 GHz) and the Measurements Laboratory of the Federal Institute of Paraíba (IFPB), Campus João Pessoa, using the VNA of Agilent Technologies model N5230A (300 kHz–13.5 GHz). The characterization of denim substrate was performed by the probe method using a VNA of Agilent model E5071C (300 kHz–20 GHz).

The rigid antennas were designed using a low-cost fiberglass laminate (FR4) as dielectric substrate with a thickness of $h = 1.5$ mm, dielectric constant of $\varepsilon_r = 4.4$, and loss tangent of 0.02.

The wearable bio-inspired antennas were designed using a denim as dielectric substrate with a thickness of h = 1 mm, dielectric constant of $\varepsilon_r = 2.14$ and loss tangent of 0.08, and flexible copper.

3.1. Wearable bio-inspired prefractal antennas

The use of wearable antennas are necessary some characteristics as easy interaction with the body, low visual impact, preferably low cost, and flexible structure [28], and for this reason, the materials used in the manufacture of the wearable antennas must follow some requirements: easy interaction with the body, flexible structure, reduced visual impact, and preferably low cost [28].

Figure 3 shows the development of wearable textile patch antenna generated by L-systems with $k = 4$ interactions. **Figure 3(a)** illustrates the original bio-inspired shape, and **Figure 3(b)**

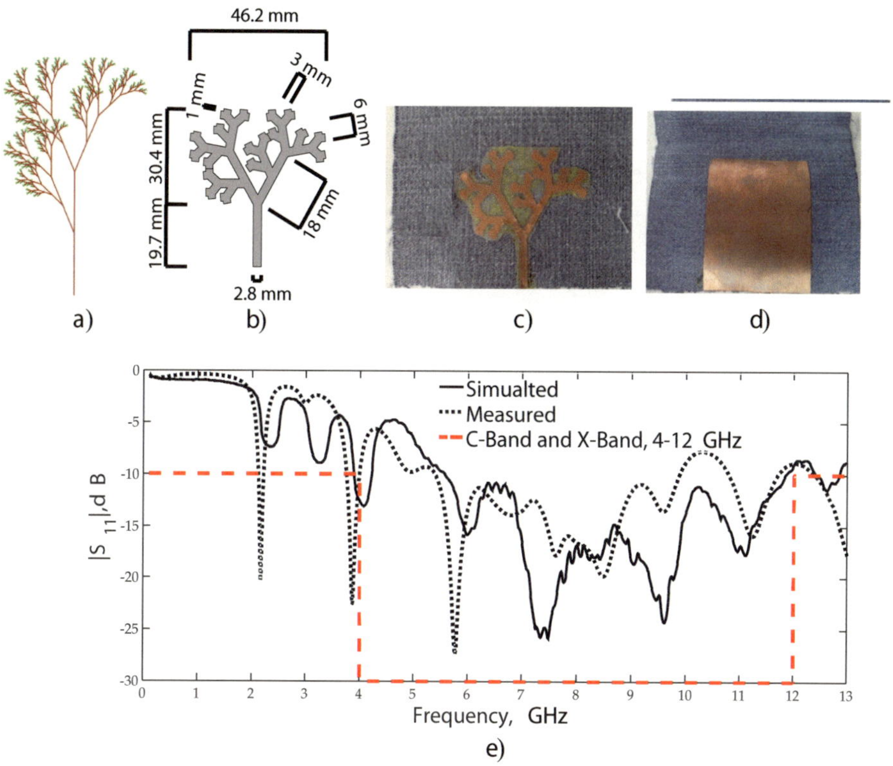

Figure 3. Development of wearable textile patch antenna generated by L-systems: (a) original shape, (b) image generated by MATLAB, (c) top layer prototype, (d) bottom layer prototype and (e) comparison between simulated and measured return loss.

presents the image generated by MATLAB® using turtle algorithm. **Figure 3(c)** and **(d)** show the top and bottom layers of the prototype built, respectively.

As observed in **Figure 3(e)**, the antenna presents an UWB behavior operating between the C-Band (4–8 GHz) and X-Band (8–12 GHz), with measured bandwidth of 5.95 GHz (5.9–1.85 GHz), and good relationship between simulated and measured results, with difference in bandwidth of 17%.

3.2. Bio-inspired polar microstrip antennas

Figure 4 shows frequency resonance of bio-inspired polar microstrip antenna for *k*-interactions (k = 1, 8 12, 16, 24, 32, 40, 48, 56, 64) and the comparison of measured $|S_{11}|$ parameter, with prototype images. The proposed of bio-inspired patch antennas is based on a circle patch antenna with displaced microstrip line feed, and quarter-wave transformer, with dimensions calculated accordingly [7, 9].

Figure 4 shows the $|S_{11}|$ parameters measured of the polar antennas to k = 8, 16, and 24. We noted that the increase of the patch perimeter by use of polar interaction provides a reduction of the resonant frequencies, similar to the fractal comportment.

Figure 5 shows the use of polar transformer in development of the array patch antenna with 4 petals, k = 8 interactions. The polar array presented good response, with simulated and measured results closed, and loss return less than −45 dB, bandwidth of 101 MHz, cover the WLAN band in 2.4 GHz.

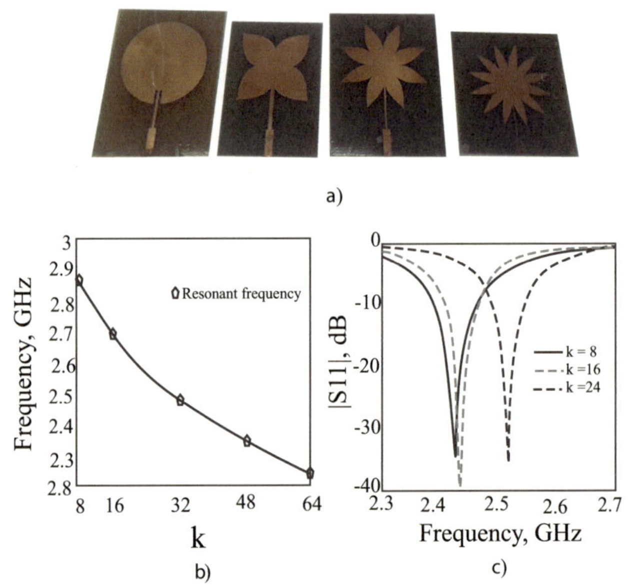

Figure 4. Interactions of bio-inspired polar microstrip patch antenna: (a) prototypes; (b) resonant frequency vs interactions and (c) return loss for k = 8, 16 and 24.

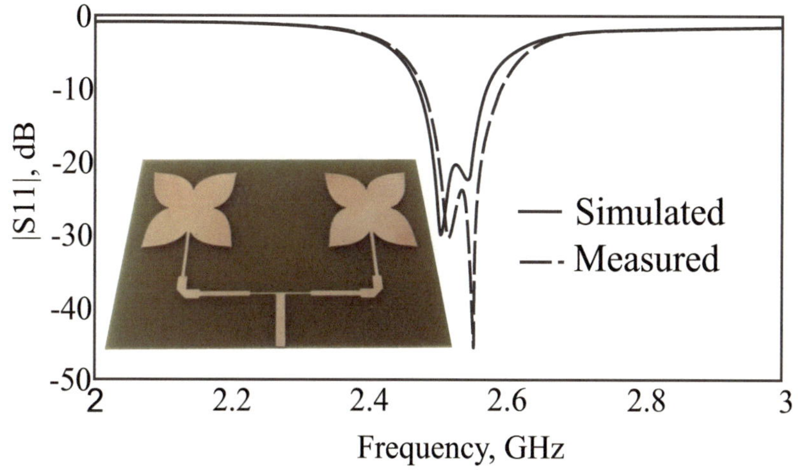

Figure 5. Polar microstrip patch antenna 1×2 array.

3.3. Leaf-shaped antennas from Gielis superformula

3.3.1. *Tulip flower–shaped antenna*

The image of the tulip flower with three petals was generated by the Gielis superformula using the software MATLAB®, with values: $m = 2$, $n_1 = 400$, n_2 and $n_3 = 1200$, a and $b = 1$. In **Figure 6** are shown the dimensions of the petals, the design simulated, and the prototype antenna.

The radiation element is composed of one central petal and two lateral petals with an inclination of 25°. The antenna was designed for a first resonance frequency at 2.1 GHz, with wavelength of $\lambda \approx 13$ mm, which was used as approximated dimension between the edge and the center of the structure.

Figure 7 presents the comparison of simulated and measured $|S_{11}|$ parameter of the tulip flower–shaped antenna. In **Table 5**, it can be observed the measured and simulated values of

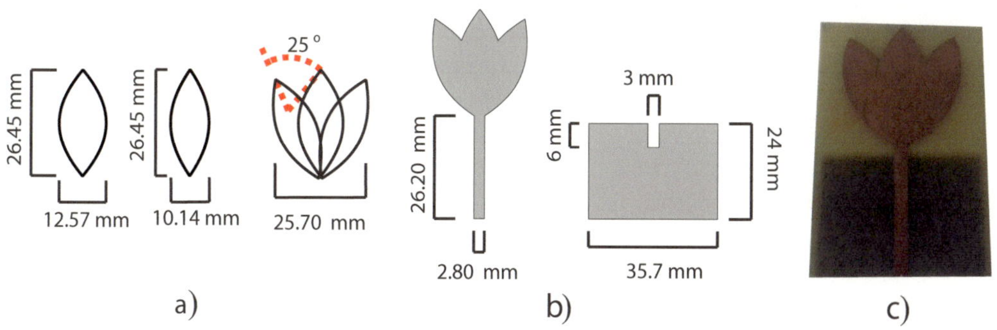

Figure 6. Tulip flower–shaped antenna for UWB applications: (a) petals exported in DXF format; (b) simulated structure top and bottom and (c) prototype.

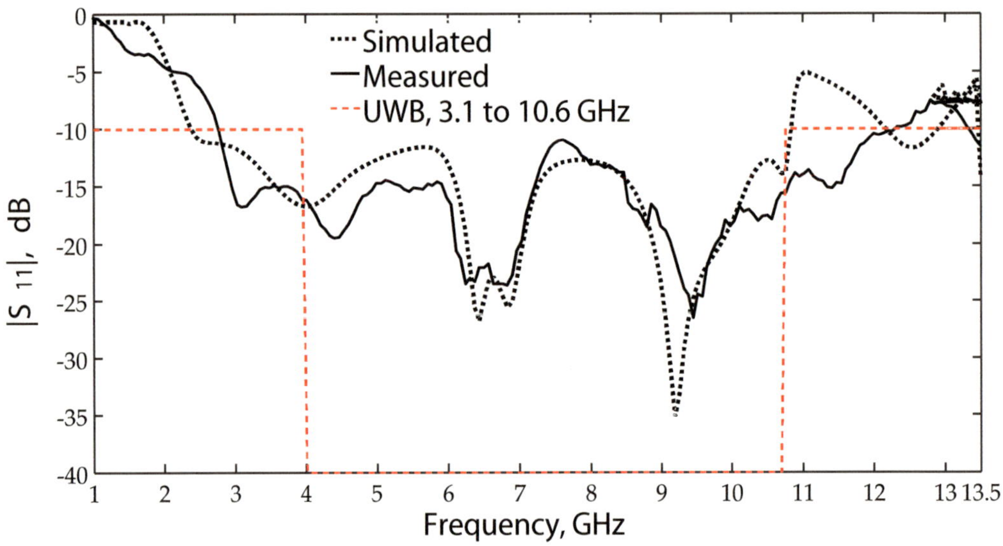

Figure 7. Comparison between simulated and measured return loss of the tulip flower–shaped antenna.

	Antenna	**BW (GHz)**	f_1 **(GHz)**	f_2 **(GHz)**
1	Simulated	9.45	1.37	10.82
2	Measured	9.56	2.77	12.33

Table 5. Measured and simulated values of resonant frequencies and bandwidth of the bio-inspired antenna.

resonant frequencies and bandwidth of the bio-inspired antenna. It can be noted that both antennas comply with the FCC parameters with a bandwidth greater than 7.5 GHz (3.1–10.6 GHz).

The 3D radiation patterns simulated for the frequencies of 6.3 and 8.6 GHz can be observed in **Figure 8(a)** and **(b)**. The 2D radiation patterns simulated for the same frequencies can be observed in **Figure 8(c)** and **(d)** with radiation pattern measured in semi-anechoic chamber for *Phi* = 90°. The results cover the FCC parameters, with omnidirectional radiation pattern, half power beam width (HPBW) greater than 60°, maximum gain in broadside direction, and current density of 6 A/m^2.

3.3.2. *Jasmine flower–shaped antenna*

Based on a circular planar monopole antenna (PMA) with a radius of 10.71 mm, a bio-inspired printed monopole antenna with the geometry of jasmine flower with 10 petals was designed. The image of the jasmine flower with 10 petals with 1 mm long was generated by Gielis superformula using the software MATLAB®, with values: $m = 10$, $n_1 = -2$, n_2 and $n_3 = 1.2$, a and $b = 1$, and the parameters: $m = 10$, $n_1 = 8$, n_2 and $n_3 = -0.6$, a and $b = 1$ for petals with 7 mm long.

http://dx.doi.org/10.5772/intechopen.69766

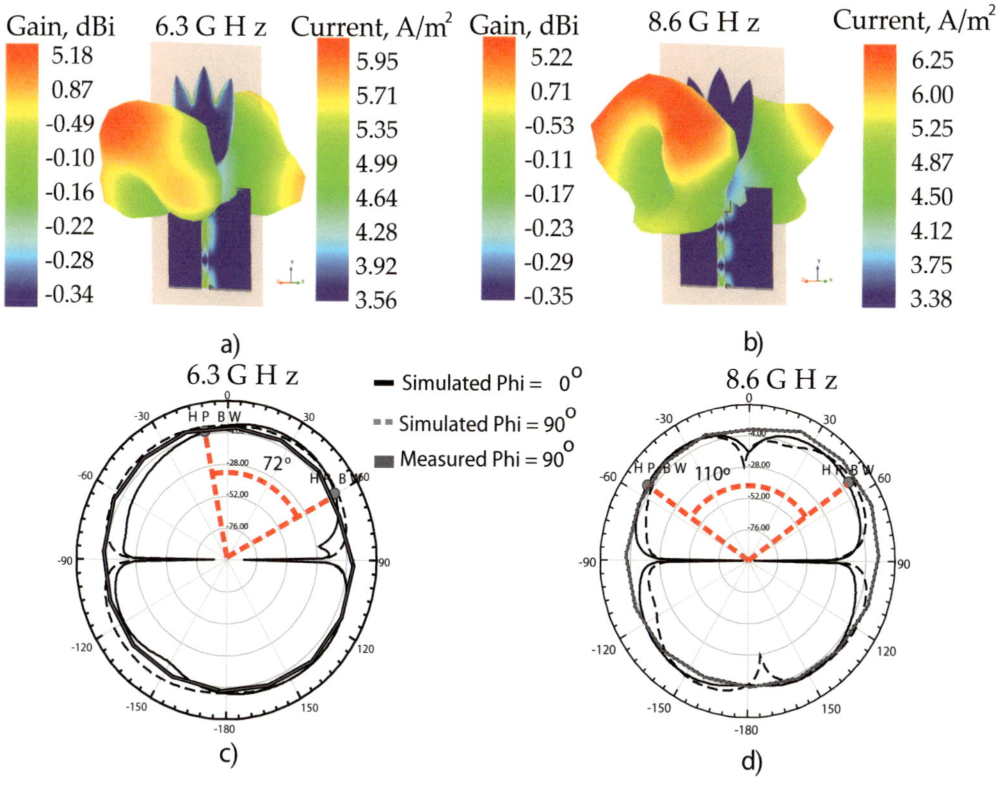

Figure 8. Radiation pattern: (a) far-field 3D with gain in dBi and current density in A/m² at 6.3 GHz; (b) far-field 3D with gain in dBi and current density in A/m² at 8.6 GHz; (c) 2D, E-plane ($\phi = 0°$) and H-plane ($\phi = 90°$) with HPBW indication at 6.3 GHz; (d) 2D, E-plane ($\phi = 0°$) and H-plane ($\phi = 90°$) with HPBW indication at 8.6 GHz.

In **Figure 9** are shown the shapes generated, simulated structures with dimensions, and prototypes of the proposed antennas. The use of the bio-inspired geometry provided a reduction of 11.3% in comparison with the classical PMA.

As seen in **Figure 10**, for the circular PMA, measured and simulated results are close, indicating convergence between the designed and built prototypes.

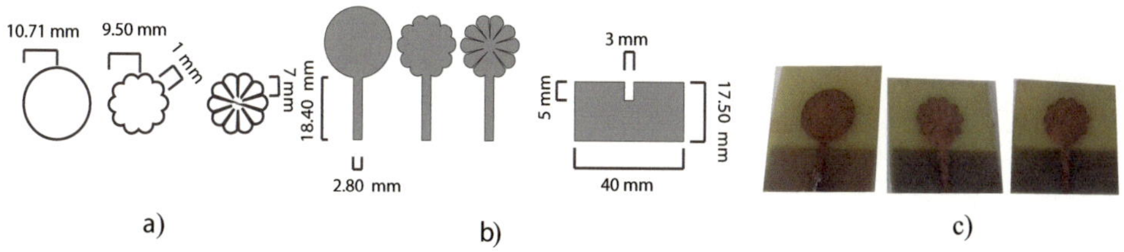

Figure 9. Jasmine flower–shaped antenna: (a) shapes exported in DXF format; (b) simulated structures, top and bottom; (c) prototypes.

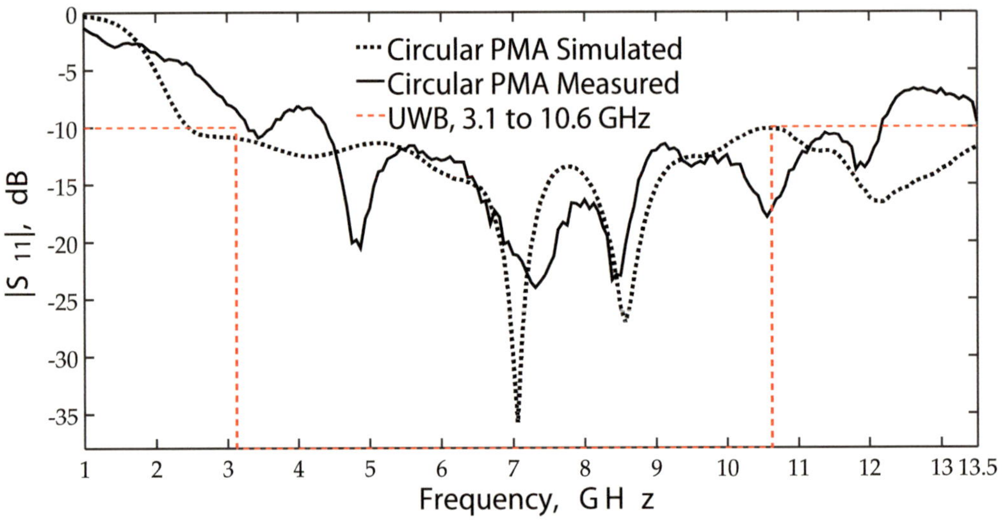

Figure 10. Comparison between simulated and measured results for designed circular printed monopole antenna.

In order to validate the simulations, two prototypes of bio-inspired PMA were built with petal lengths of 1 and 7 mm. The comparisons between simulated and measured results are shown in **Figures 11** and **12**.

It can be seen, from the **Figures 11** and **12**, that increasing the length defining the petals allows a reduction in the radius of the patch element in 11.30% and promotes the modification of the resonance frequency. The antenna with 7 mm presented an operating range within the range of the X-Band frequency. The measurements of the return loss of the circular PMA and the bio-inspired antenna with a petal length of 7 mm present a higher bandwidth than the simulation.

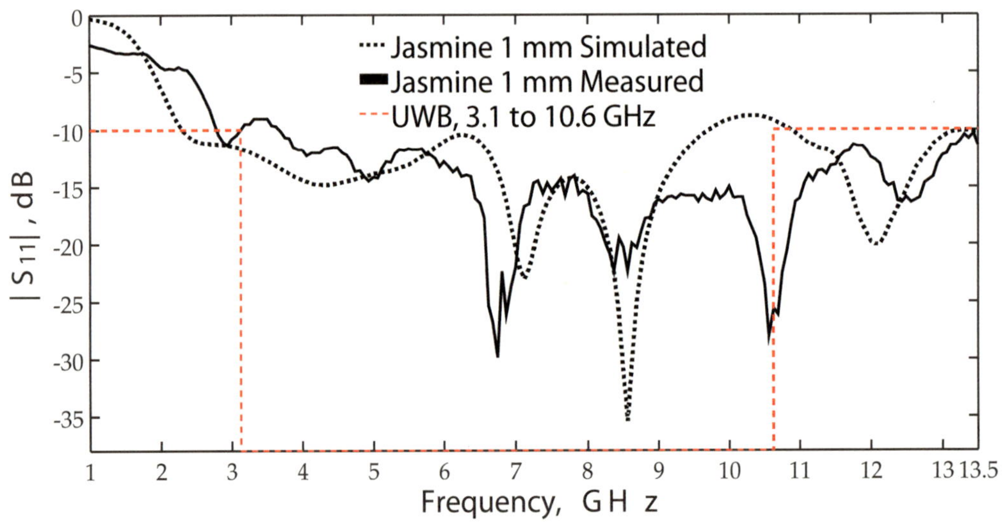

Figure 11. Comparison between simulated and measured results for jasmine flower PMA with petal length of 1 mm.

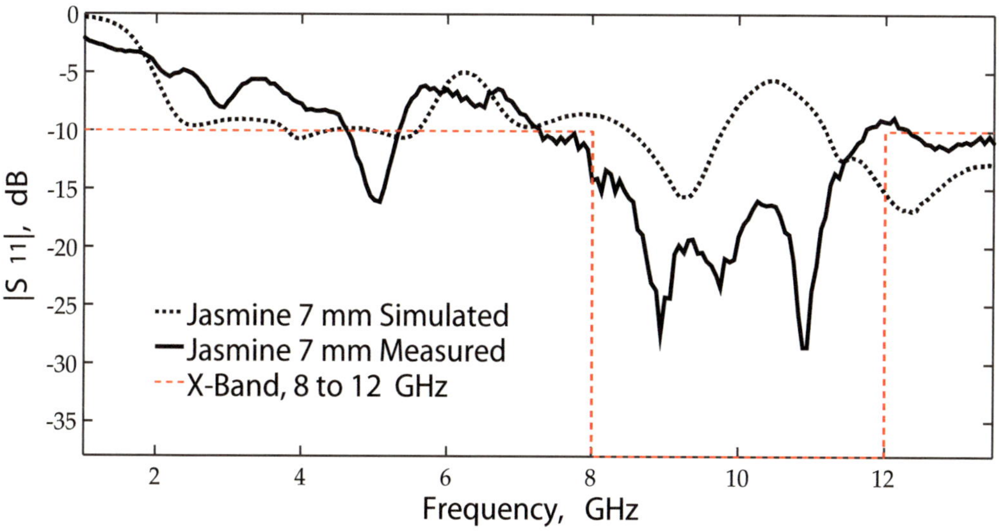

Figure 12. Comparison between simulated and measured results for jasmine flower PMA with petal length of 7 mm.

The bio-inspired antenna with a petal length of 1 mm presents a bandwidth 11% lower compared to simulations, but still cover the UWB range.

The bio-inspired geometry of a jasmine flower increases the perimeter of the antenna compared to a classical PMA and consequently the frequency behavior. The perimeter of the antenna is bigger without changing its size. In the case of petal, length of 7 mm is possible to change the resonance frequency and bandwidth to operate in the X-band range (8–12 GHz).

The **Figure 13** shows the measured values of the circular PMA and the bio-inspired antennas with petal length of 1 and 7 mm.

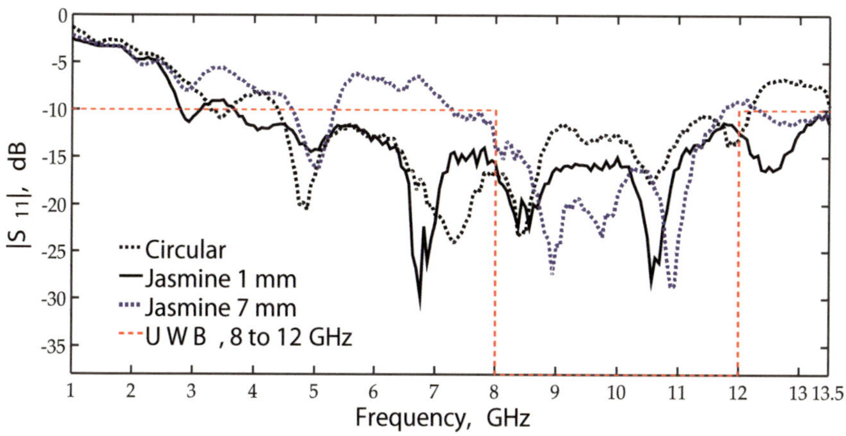

Figure 13. Comparison between measurements of $|S_{11}|$ for built printed monopole antennas: circular and jasmine flowers (1 and 7 mm).

The difference between the resonance frequencies and bandwidth of the built antennas is presented in **Table 6**.

Figure 14 shows the 3D simulated radiation patterns of the circular PMA and the bio-inspired antennas with petal length of 1 mm and 7 mm. The radiation patterns at 7.13 GHz of the circular PMA, at 7.06 GHz, of the jasmine flower with petal length of 1 mm and at 9.25 GHz of the jasmine flower with petal length of 7 mm are presented in **Figure 14(a)–(c)**, respectively. The antennas presented maximum gain in the broadside direction close to 6 dBi and omnidirectional radiation pattern.

3.3.3. Wearable ginkgo biloba leaf-shaped antenna

The ginkgo biloba is a plant of Chinese origin; its name means silver apricot and can be found on all continents. The leaves of this plant have interesting characteristics, such as good ratio between its length and width, and a geometry with a larger perimeter than Euclidean geometries. Those characteristics, in a PMA, provide a broadband antenna, with reduced dimensions, covering the lower frequency bands. This feature allows a compact antenna design and the

	Antenna	**BW (GHz)**	**f_1 (GHz)**	**f_2 (GHz)**
1	Circular simulated	7.38	2.31	9.69
2	Circular measured	10.69	2.81	13.50
3	Jasmine 7 mm simulated	11.00	2.50	13.50
4	Jasmine 7 mm measured	8.45	3.75	12.20
5	Jasmine 1 mm simulated	4.25	8.00	12.25
6	Jasmine 1 mm measured	3.44	9.25	11.69

Table 6. Comparison of frequency response and bandwidth.

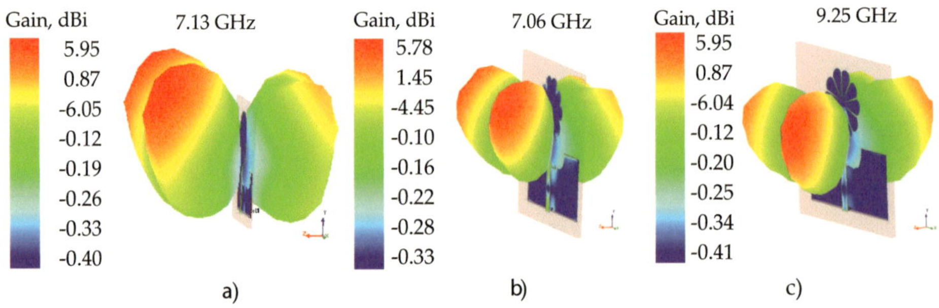

Figure 14. 3D simulated radiation patterns: (a) PMA at 7.13 GHz, (b) 1 mm jasmine flower at 7.06 GHz, (c) 7 mm jasmine flower at 9.25 GHz.

possibility of use in different frequency bands. From the determination of the perimeter, the use of geometry of ginkgo biloba leaf was applied to a broadband antenna structure, in which it was possible to run the project for an antenna that covers the frequencies of the technologies 2G, 3G and 4G.

The image of the ginkgo biloba leaf used for the proposed antenna was generated using the software MATLAB®, with values: $m = 4$, $n_1 = -0.1$, $n_2 = 0.14$, $a = 0.1$ and $b = 1$.

In **Figure 15** are shown the shape generated and the simulated structure with dimensions and, in **Figure 16**, the prototype of the proposed antenna.

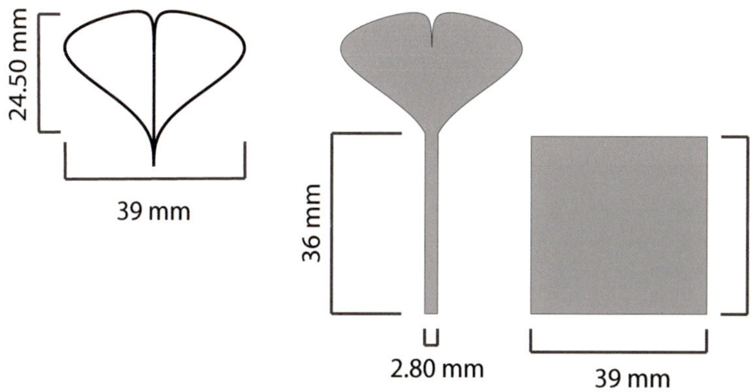

Figure 15. Simulated structure of the textile PMA ginkgo biloba.

Figure 16. Textile PMA ginkgo biloba prototype.

In **Figure 17** is shown the comparison of the values of the parameter $|S_{11}|$ measured and simulated of the PMA ginkgo biloba. As noted, the measurement and simulation show similar results, indicating convergence between the simulation and the prototype.

In **Figure 18**, the parameter $|S_{11}|$ measured of the PMA ginkgo biloba in the pocket can be seen, close of the head and hand at distance of 20 mm. As noted, the antenna used close to the body suffers interference, which promotes the shift of the frequency and bandwidth of the antenna. In **Table 7**, the measured and simulated values, resonant frequencies, and bandwidth of the bio-inspired antenna can be observed. The resonance frequencies have difference <4,2%, the first frequency and bandwidth covering 2G (1850–1900 MHz), 3G (1920–1975 MHz) and 4G (LTE; 2500–2690 MHz) bands.

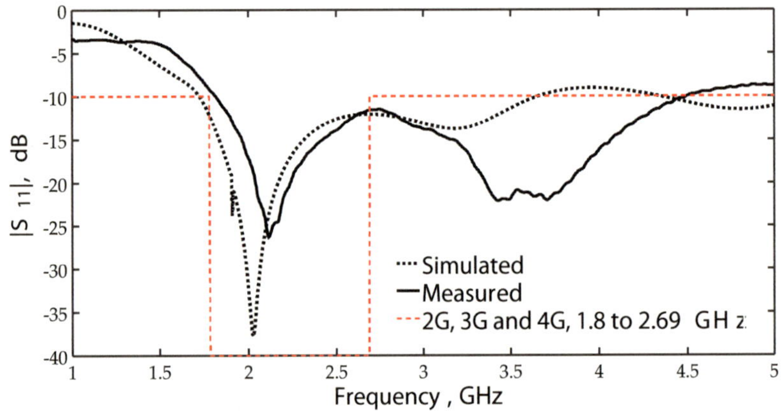

Figure 17. Simulated and measured $|S_{11}|$ parameter of the PMA ginkgo biloba.

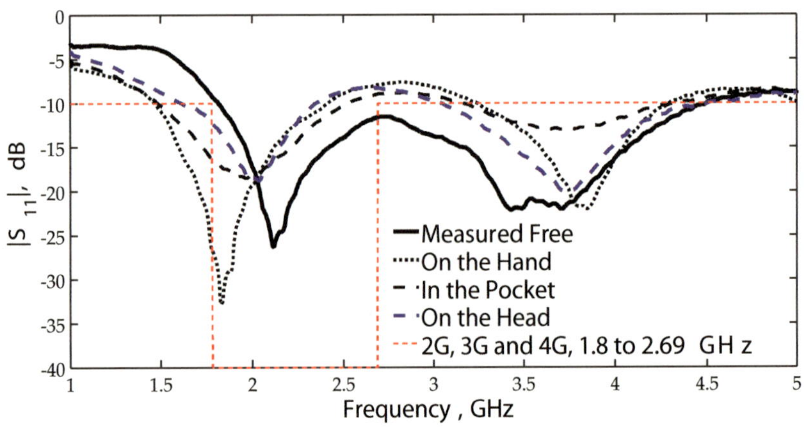

Figure 18. Simulated and measured $|S_{11}|$ parameter of the PMA ginkgo biloba with body interferences.

http://dx.doi.org/10.5772/intechopen.69766

	Antenna	BW (GHz)	f_1 (GHz)	f_2 (GHz)	f_0 (GHz)
1	Simulated	3.51	1.73	5.24	2.03
2	Measured	2.70	1.80	4.50	2.12
3	Measured on head	0.73	1.61	2.34	2.04
4	Measured in pocket	1.04	1.49	2.53	1.97
5	Measured on hand	0.94	1.47	2.41	1.83

Table 7. Comparison of frequency response and bandwidth of the simulated and measured PMA ginkgo biloba.

The measurements, using the antenna nearly the head, hand, and in the pocket, showed different resonant frequencies and bandwidths. The biggest difference in the resonance frequency was observed nearly the hand, with a difference of 13.68%. The biggest difference in bandwidth was detected at the antenna close to the head, with a difference of 72.96%.

The **Figure 19** shows the 2D and 3D radiation patterns of the PMA textile ginkgo biloba at 2.12 GHz, with a maximum gain 3.16 dBi in the broadside direction. The 2D radiation pattern measured for *Phi* = 90° was performed in semi-anechoic chamber. It can be observed that the half-power beam width (HPBW) in (ϕ = 90) E-plane is ~120° with an omnidirectional radiation pattern. By presenting equivalent power distribution and omnidirectional radiation diagram, this antenna can be used in applications that require direct communication between devices, in which the transmitter/receiver can take various positions.

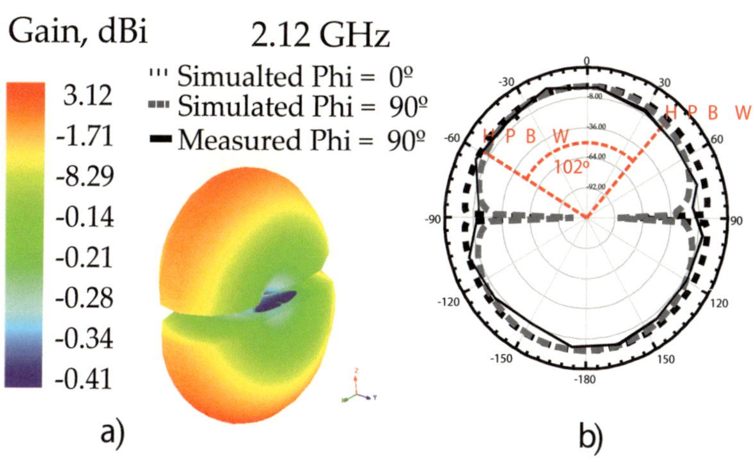

Figure 19. 2D and 3D radiation pattern of the antenna proposed.

3.3.4. Wearable Bidens pilosa leaf-shaped antenna

The Bidens pilosa is native to the Americas, but it is known widely as an introduced species of other regions. The use of geometry of Bidens pilosa with three leaves was applied to a broadband antenna structure, in which it was possible to run the project for an antenna that covers the WLAN range at 2.40 GHz (2.40–2.4835 GHz).

Based on the perimeter of a classical circular antenna, it is possible to design the bio-inspired textile patch antenna with three elliptical leaves using the Gielis formula. Thus, the leaves are generated by the parameters $n_1 = 2$, $m = 400$, n_2 and $n_3 - 1200$, a and $b = 1$. The final structure obtained total perimeter of 143.3 mm.

In **Figure 20** are shown the circular patch simulated, the dimensions of the single leaf used, the simulated dimensions of the bio-inspired array antenna and the prototype, respectively. In the simulation and the prototype, the under leaf was inclined at 20°, and the down leaves were inclined at 40° in relation to the geometry of Bidens pilosa, in order to provide fine-tuning of the resonance frequency.

Figure 21 shows a comparison between the simulated $|S_{11}|$ parameter for the circular patch antenna and the bio-inspired textile antenna. The circular patch antenna obtains a bandwidth of 140 MHz and the bio-inspired wearable patch antenna of 130 MHz. Both antennas have bandwidth that fully covers the required frequency range for WLAN technology. However, the bio-inspired array patch antenna presents a reduction of total length in 33.61% and width in 52.01% compared to the circular patch antenna.

Figure 22 shows a comparison between the simulated and measured $|S_{11}|$ parameter for the bio-inspired wearable textile patch antenna proposed. As it can be observed, the simulated and measured prototypes show a convergent behavior. The measured bandwidth is 8.33% narrower than the simulated values, but still fully covering the WLAN band.

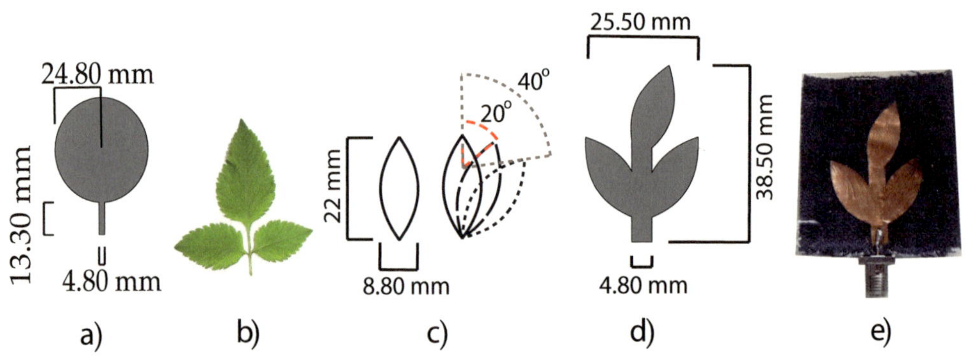

Figure 20. (a) Circular patch simulated (b) leaves array of *Bidens pilosa*, (c) single elliptical leaf, (d) bio-inspired patch antenna simulated, (e) bio-inspired prototype.

http://dx.doi.org/10.5772/intechopen.69766

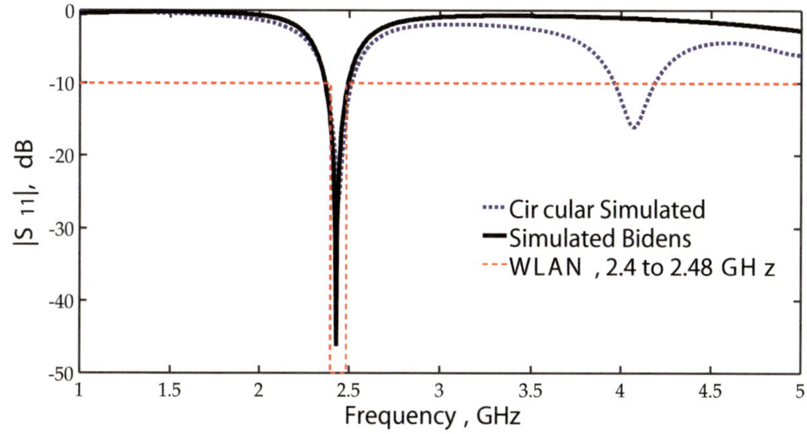

Figure 21. Comparison between $|S_{11}|$ parameter of the circular patch and the bio-inspired textile antenna.

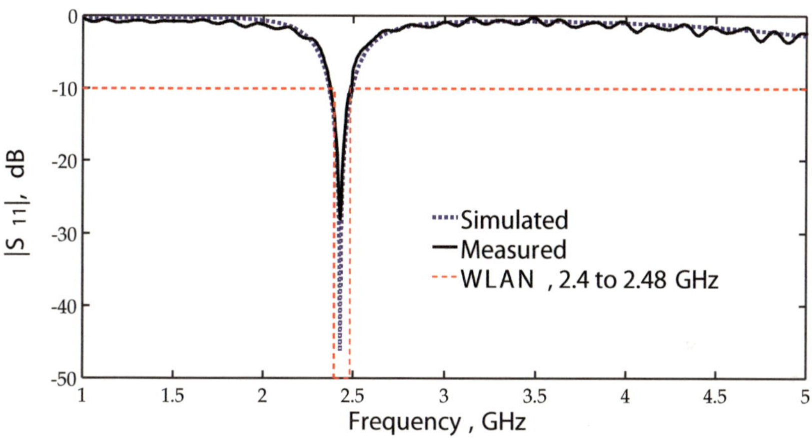

Figure 22. Comparison of $|S_{11}|$ parameter simulated and measured of *Bidens pilosa* patch antenna.

In **Figure 23** is illustrated the radiation pattern in 3D and 2D with the half power beam width and relative front-to-back (F/B) for the simulated bio-inspired textile array antenna. The 2D radiation pattern measured for *Phi* = 90° was performed in semi-anechoic chamber. It can be noted that the antenna presents an end-fire direction with maximum gain of 6.71 dBi, HPBW = 90° (indicated by the letters 'A' and 'B') and F/B = 21 dB.

In **Figure 24** is shown the simulated current density of the bio-inspired textile antenna and the circular antenna at 2.43 GHz. It can be observed that the surface current is more distributed on the edges than on the center of the patch. The current density of the circular patch is 5.86 A/m^2, and the bio-inspired is 40.06 A/m^2. Therefore, the bio-inspired antenna presents a higher concentration of the surface current in a smaller physical area of the antenna.

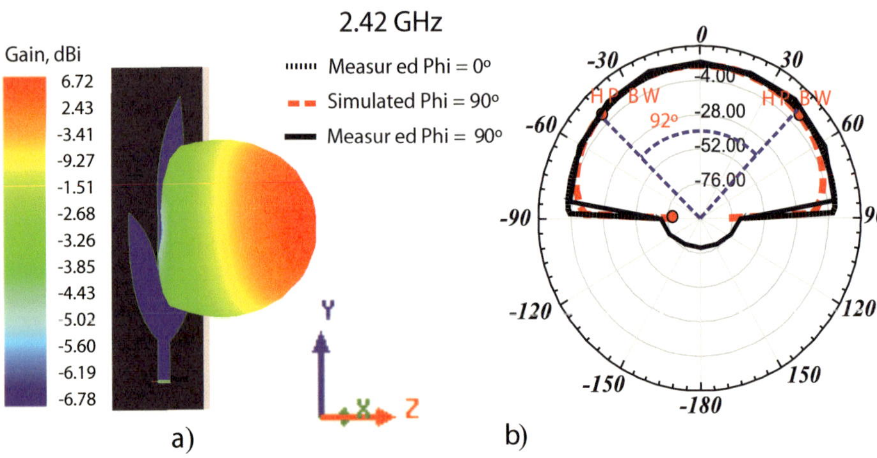

Figure 23. Radiation pattern of the bio-inspired textile antenna: (a) 3D with antenna structure, (b) 2D with HPBW and relative F/B.

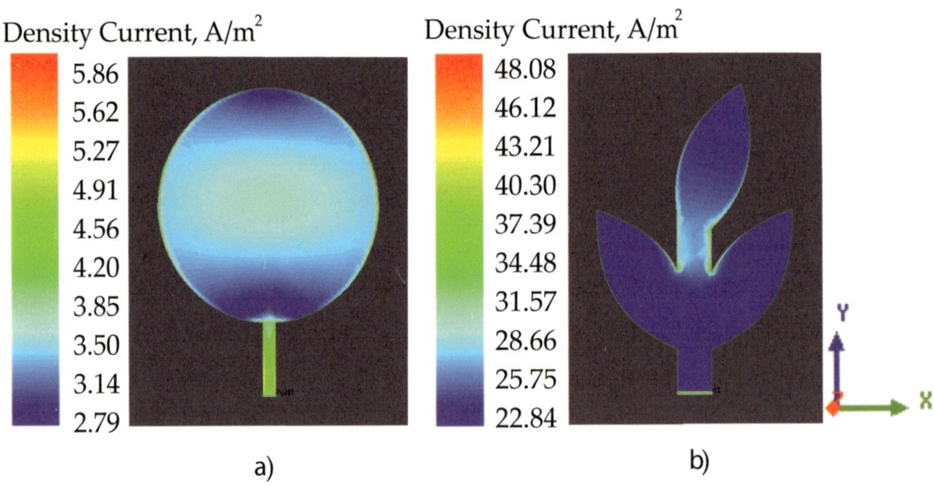

Figure 24. Current density distribution: (a) circular patch antenna, (b) bio-inspired textile patch antenna.

4. Conclusion

In this chapter, we have described some trends for design innovative bio-inspired microstrip antennas. The methods of analysis, manufacturing, and measurement have presented considering different dielectric materials (rigid and flexible) for the manufacture of the antennas. The unique properties of space filling and self similarity naturally result in more compact and multiband behavior antennas. Consequently, these antennas have their gain reduced, which in many wireless applications are undesirable characteristics. On the other hand, as observed on the last case, the bio-inspired antenna presents a higher concentration of the surface current in a smaller physical area compared to classical geometries with esthetic appeal. The bio-inspired microstrip antennas have few design variables and smooth responses in the region of interest of these design variables, which facilitates all steps of the design methodology. These characteristics open a vast research field for wearable embedded antennas.

Author details

Alexandre Jean René Serres[1]*, Georgina Karla de Freitas Serres[1], Paulo Fernandes da Silva Júnior[1], Raimundo Carlos Silvério Freire[1], Josiel do Nascimento Cruz[1], Tulio Chaves de Albuquerque[1], Maciel Alves Oliveira[1] and Paulo Henrique da Fonseca Silva[2]

*Address all correspondence to: alexandreserres@gmail.com

1 Electrical Engineering – COPELE, Federal University of Campina Grande – UFCG, Campina Grande, Brazil

2 Electrical Engineering, Federal Institute of Paraiba – IFPB, Paraiba, Brazil

References

[1] R.A. Sainati, Cad of microstrip antennas for wireless applications, Artech House, Norwood, MA, USA, 1996.

[2] Z. N. Chen, Antennas for Portable Devices. Chichaster: John Wiley & Sons, 2007. 290 p.

[3] Z. N. Chen and M.Y.W. Chia, Broadband planar antennas: design and applications. John Wiley & Sons, Ltd, Chichaster, England, 2006.

[4] K.-L. Wong, Compact and broadband microstrip antennas, John Wiley & Sons, Inc., New York, USA, 2002.

[5] 5G Americas, 5G Americas White Paper on 5G Spectrum Recommendations- April 2017, [Online]. Available: http://www.5gamericas.org/files/9114/9324/1786/5GA_5G_Spectrum_Recommendations_2017_FINAL.pdf. [Accessed: 12- Jun- 2017].

[6] C. A. Balanis. Antenna Theory. 3rd ed. Arizona: Wiley; 2009. 941 p.

[7] R. Garg, P. Bhartia, I. Bahl, and A. Ittipiboon, Microstrip antenna design handbook, Artech House, Boston, USA, 2001.

[8] W. L. Stutzman, Antenna Theory and Design, 2 ed. New York: John Wiley & Sons, 1998. 598 p.

[9] G. Kumar and K. P. Ray, Broadband microstrip antennas, Artech House, Boston, Mass, USA, 2003.

[10] Yoseph Bar-Cohen. Biomimetics: Biologically Inspired Technologies. Boca Raton, FL: Taylor and Francis; 2006

[11] Falko Dressler and Ozgur B. Akan. A survey on bio-inspired networking. Computer Networks. 2010;**54**(6):881-900. DOI: 10.1016/j.comnet.2009.10.024

[12] J.A. Flint. A biomimetic antenna in the shape of a bat's ear. IEEE Antennas and Wireless Propagation Letters. 2006;**5**(1):145-147. DOI: 10.1109/LAWP.2006.873940

[13] Nader Behdad, Mudar A. Al-Joumayly and Meng Li. Biologically Inspired Electrically Small Antenna Arrays With Enhanced Directional Sensitivity. IEEE Antennas and Wireless Propagation Letters . 2011;**10**:361-364. DOI: 10.1109/LAWP.2011.2146223

[14] Khabat Ebnabbasi. A Bio-Inspired Printed-Antenna Transmission-Range Detection System. IEEE Antennas and Propagation Magazine. 2013;**55**(3):193-200. DOI: 10.1109/MAP.2013.6586661

[15] G.H. Huff. Biologically-inspired vascular antenna reconfiguration mechanism. Electronics Letters. 2011;**47**(11):637 - 638. DOI: 10.1049/el.2011.0383

[16] Xiao-Feng Bai, Shun-Shi Zhong and Xian-Ling Liang. Leaf-shaped monopole antenna with extremely wide bandwidth. Microwave and Optical Technology Letters. 2006; **48**(7):1247-1250. DOI: 10.1002/mop.21668

[17] Soh Fujita, Manabu Yamamoto and Toshio Nojima. A study of a leaf-shaped bowtie slot antenna for UWB applications. In: IEEE, editor. International Symposium on Antennas and Propagation (ISAP); 29 Oct.-2 Nov. 2012;Nagoys: p. 830-833.

[18] A. N. Askarpour, A. Gholipour and R. Faraji-Dana. A Band-Notched Tulip Antenna for UWB Applications. In: 38th European Microwave Conference; 27-31 Oct. 2008; Amsterdam: IEEE; 2008. p. 881-884. DOI: 10.1109/EUMC.2008.4751594

[19] F. M. Tanyer-Tigrek, D. P. Tran, I. E. Lager and L. P. Ligthart. Wide-band tulip-loop antenna. In: 3rd European Conference on Antennas and Propagation; 23-27 March 2009; Berlin: IEEE; 2009. p. 1446-1449.

[20] S. R. Patre and S. P. Singh. MIMO antenna using castor leaf-shaped quasi-self-complementary elements for broadband applications. In: IEEE MTT-S International Microwave and RF Conference (IMaRC); 10-12 Dec. 2015; Hyderabad: IEEE; 2015. p. 140-142. DOI: 10.1109/IMaRC.2015.7411400

[21] P. F. Silva Júnior, P. H. da F. Silva, A. J. R. Serres, J. C. Silva and R. C. S. Freire. Bio-inspired design of directional leaf-shaped printed monopole antennas for 4G 700 MHz band. Microwave and Optical Technology Letters. 2016;**58**(7):1529-1533. DOI: 10.1002/mop.29853

[22] D. M. Sullivan, Electromagnetic Simulation Using the FDTD Method. New York: Wiley: IEEE Press, 2000.

[23] MANDELBROT, B. B. The fractal geometry of nature. 3. ed. Nova York: W. H. Freeman and Co., 1982. 468p.

[24] FALCONER, K. Fractal geometry: mathematical foundations and application. 2. ed. Londres: Wiley, 2003. 337p.

[25] Mishra, J. and Mishra, S. L-Systems Fractals. Amsterdam, Netherlands: Elsevier, 2007. 274p.

[26] Barnsley, M. Fractals Everywhere. San Diego: Academic Press, 1988. 394 p.

[27] Johan Gielis. A generic geometric transformation that unifies a wide range of natural and abstract shapes. American Journal of Botany. 2003;**90**(3):333-338. DOI: 10.3732/ajb.90.3.333

[28] E. E. C. Oliveira, P. H. da F. Silva, A. L. P. S. Campos, S. G. Silva. Overall size antenna reduction using fractal elements. Microwave and Opt Technol Letters. 2009;**51**(3):671-675. DOI: 10.1002/mop

Printed Planar Antenna Designs Based on Metamaterial Unit-Cells for Broadband Wireless Communication Systems

Mohammad Alibakhshikenari,
Mohammad Naser-Moghadasi,
Ramazan Ali Sadeghzadeh, Bal Singh Virdee and
Ernesto Limiti

Additional information is available at the end of the chapter

http://dx.doi.org/10.5772/intechopen.68600

Abstract

With the continuing development of mobile communications, the communication standards, which include operating frequencies and protocols, are also evolving. In order to accommodate these and future changes, antennas with characteristics of wideband and multiband are becoming a necessity. Hence, wireless communications industries are now demanding broadband antennas that are low-profile and low-volume structures. Conventional planar microstrip antennas are the most common form of printed antennas that have been used for many years. This is because these antennas offer advantages of low cost, conformability, and ease of manufacturing; however, the bandwidth of these types of antennas is highly restricted. Among different types of planar antennas, the slotted structure that offers the simplest structure is compact and radiates omnidirectionally; these features make it an excellent candidate for broadband applications.

Keywords: planar slotted antennas, metamaterials, composite right-/left-handed transmission lines, traveling-wave antenna, wireless communications systems

1. Introduction

Printed planar antennas such as patch antennas are favored over other types of antennas as they are low profile, light weight, and can be easily manufactured using low-cost PCB techniques. These types of antennas are constructed using microstrip integrated circuit technology, where a metal layer is separated from the ground-plane by a dielectric substrate [1, 2]. The thickness of the metal layer is much smaller than the operating wavelength. Such antennas are conventionally excited through a feed-line with a voltage between the patch and the ground plane. This excites a current on the patch, and hence, a vertical electric field between the patch and the ground plane [3]. The patch element resonates when its width or diameter is near $\lambda_g/2$, leading to relatively large current and field amplitudes, and the radiation mechanism occurs from the fringing fields between the edge of the microstrip and the ground plane. Printed planar antennas essentially radiate energy in a direction perpendicular to the plane of the antenna substrate with moderately high gain [3–5].

The choice of the substrate used is important (dielectric constant, thickness, and loss tangent). There are various shapes of patch antennas (circular being the most common) [4]; however, the principles of operation are essentially the same. Circular patches offer advantages over other geometries for applications such as arrays and can also be easily modified to produce a range of impedance values and radiation patterns.

Microstrip patch antennas enjoy many advantages when compared to other types of antennas; they are relatively small, low profile, low cost, light weight, and can be integrated to other circuits or active devices which can possibly result in a single-board solution, unlike wire, waveguide, or horn antennas for example. This also allows this type of antenna to be easily mounted on objects, such as computers or flying bodies, and can be easily employed for array design.

The most common disadvantage of microstrip antennas is the inherent narrow impedance bandwidth, which is due to the thickness, which is normally thin. The easiest and most well-known method of moderately improving the bandwidth is by increasing the height of the substrate [4], which has the effect of increasing the radiation conductance. However, this is not a desirable option as it leads to introducing surface waves and power loss, but usually works well for bandwidths up to 4–6% [6]. Other disadvantages of microstrip patch patch antenna are low efficiency, low power, and poor polarization purity [5, 7]. To overcome this drawback, various metamaterial (MTM) structures, including split-ring resonator (SRR) [8], spiral, rod, omega, S-shape, and symmetric rings, etc., have been explored in literature [9, 10].

This chapter presents the design and measured performance of compact planar slotted antennas based on unique MTM structures for application in broadband wireless communication systems. Typical realization of MTM transmission line (TL) is based on quasi-lumped TL with elementary cells consisting of a series capacitor and a shunt inductor. In reality, the loss associated with the MTM structure introduces series parasitic inductance in the capacitor, and parasitic shunt capacitance in the inductor; thus, the resulting MTM structure is more accurately represented by a composite right/left-handed (CRLH) transmission-line structure. The optimized "d-shaped" antenna presented in the chapter is shown to operate over a wideband from 0.75 to 4.5 GHz with a peak gain of 3.5 dBi and efficiency of 60% at 2 GHz. Moreover, the antenna is compact and easy to fabricate using standard PCB techniques. In addition, an "X-shaped" antenna, which is also presented here, is shown to operate from 0.4 to 4.7 GHz with a gain of 2 dBi and radiates energy with an efficiency of 65% at 2.5 GHz.

http://dx.doi.org/10.5772/intechopen.68600

2. Metamaterial traveling-wave antenna

In this section, a traveling-wave planar antenna is implemented using composite right/left-handed structures. The antenna is constructed using a unique metamaterial unit cell implemented by etching "X-shaped" slots inside a rectangular patch. The patch is inductively terminated to the ground plane through a via-hole, as shown in **Figure 1**. The proposed antenna is henceforth referred to as "X-shaped" antenna. The resulting structure displays negative permittivity and permeability characteristics over a given frequency range. The aperture of the antenna can be increased by simply cascading together a number of these metamaterial unit cells, as shown in **Figure 1**. Hence, the antenna's gain and radiation efficiency are improved with negligible effect on its fractional bandwidth, which is contrary to what is observed in traditional antennas.

The antenna was designed and fabricated on Rogers RT/Duroid®5880 substrate, which is a glass microfiber reinforced PTFE composite with dissipation factor $\tan\delta = 0.0009$, dielectric constant $\varepsilon_r = 2.2$, and height $h = 1.6$ mm. The proposed antenna structure essentially acts as a guiding structure for traveling waves; thus, the traveling wave forms the main radiating mechanism. The antenna is excited on the left-hand side at port#1, and is terminated on the right-hand side at port#2 with a matched 50 Ω load, i.e., SMD1206, which is terminated to ground. Surface currents associated with the RF signal travel over the antenna structure in one direction from port#1 to the termination port#2. This is in contrast to conventional standing wave or resonant antennas, such as the monopole or dipole antennas. The nonresonating traveling-wave antenna, shown in **Figure 1**, exhibits a wide operating bandwidth.

The equivalent electrical circuit representing the metamaterial unit cell is shown in **Figure 2(a)**. In the unit cell, the "X-shaped" slot essentially behaves like a series left-handed (LH) capacitance (C_L), and the high impedance stub acts like a shunt LH inductor (L_L). Unavoidable parasitic effects in the unit cell are generated by gaps between patch and ground plane and the current flows over the patch. The parasitic anomalies are represented by the shunt right-handed (RH) capacitance (C_R) and the series RH inductance (L_R). Loss in the structure is modeled by the resistance (R_R, R_L) and conductance (G_R, G_L). Components R_R and G_R represent right-handed loss, and R_L and G_L represent left-handed loss. Important characteristics of the metamaterial unit cell were obtained from the analysis of the loss-less equivalent circuit as described in Ref. [10]. At low frequencies, L_R and C_R can be considered short-circuited and open-circuited, respectively, so that the equivalent circuit can be essentially reduced to a series-C_L/shunt-L_L circuit, which is left-handed since it has antiparallel phase and group velocities. The resulting left-handed circuit has a high-pass nature; therefore, a left-handed stopband is presented below a certain cut-off frequency. At high frequencies, C_L and L_L can be considered as short-circuited and open-circuited, respectively, so that the equivalent circuit can be reduced to the series-L_R/shunt-C_R circuit, which is right-handed since it has parallel phase and group velocities. This left-handed circuit has a low-pass nature; therefore, a right-handed stopband is presented above a certain cut-off frequency. The analysis in Ref. [10] shows that the left-handed structure exhibits negative permeability and permittivity. Unlike conventional or normal right-handed unit cell-based structures, the distinguishing feature of a LH structure is its dimensions; in particular, its length is independent of the guided wavelength, which makes the structure significantly smaller.

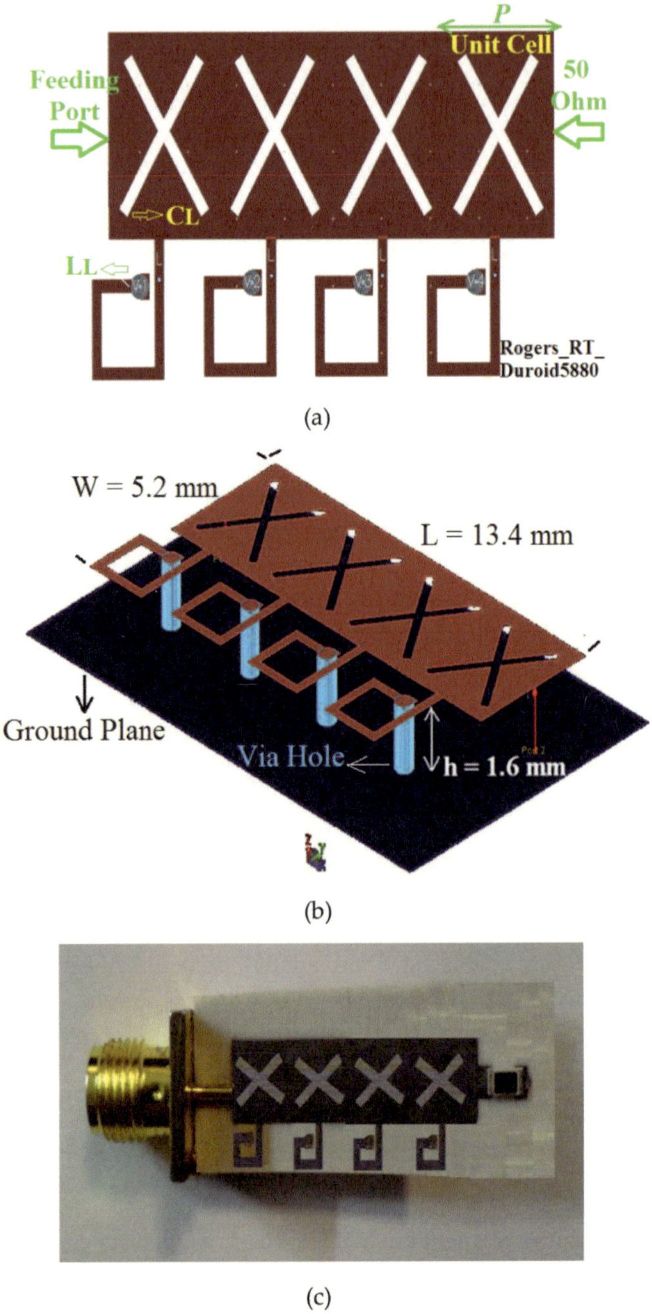

Figure 1. Antenna layout, (a) top layer, (b) isometric view of the simulation model, and (c) the fabricated prototype.

Dispersion diagram of the metamaterial unit cell structure, shown in **Figure 2(b)**, was obtained by substituting the S-parameters, which were obtained with ANSYS HFSS™ (high frequency structure simulator), in Eq.(1) as described in Ref. [10]:

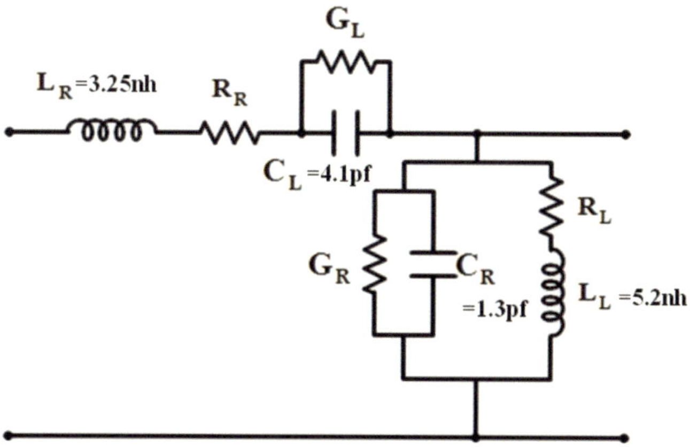

(a)

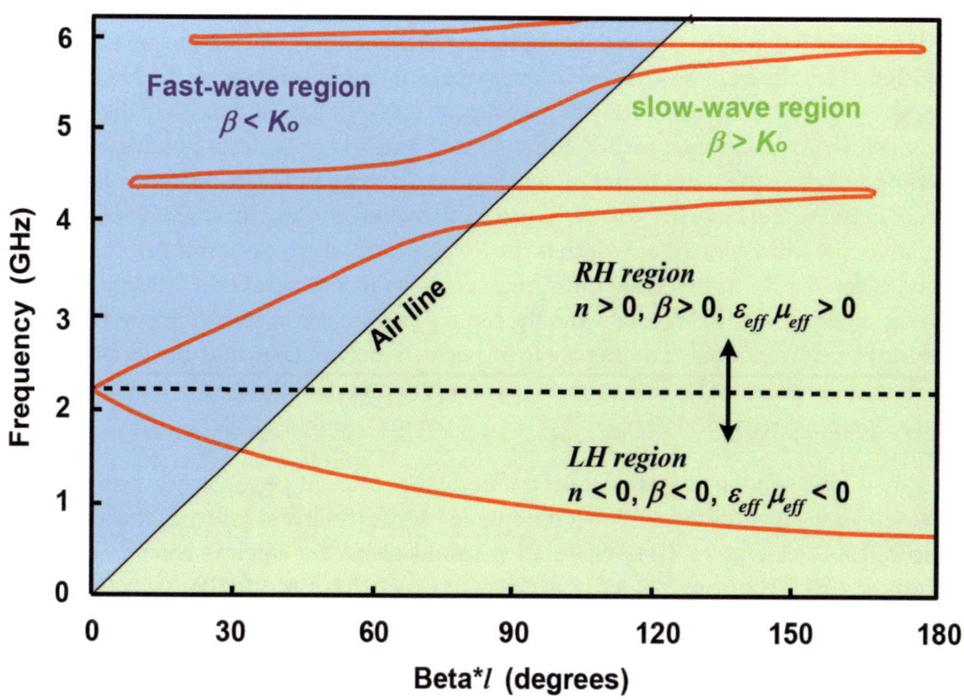

(b)

Figure 2. (a) Equivalent circuit model of the unit-cell and (b) the dispersion diagram of the loss-less CRLH unit cell.

$$\beta = \cos^{-1}\left(\frac{1 - S_{11}S_{12} + S_{12}S_{21}}{2S_{21}}\right)/l \tag{1}$$

The dispersion diagram shows the range of left-handed characteristics (i.e., 0.65–2.25 GHz) and the range of right-handed characteristics (i.e., 2.25–4.4 GHz). The slow-wave and fast-wave regions have been indicated in **Figure 2(b)**. The expression of Air-line, which separates the fast-wave ($b < K_o$) and slow-wave ($b > K_o$) regions (where K_o is the free-space wave number) is given by:

$$\text{Air-line} = \frac{\omega p}{c} \tag{2}$$

where p is the period of the unit cell as shown in **Figure 1**, and c is the speed of light.

The antenna's characterizing parameters are: $L = 13.4$ mm, $W = 5.2$ mm, $h = 1.6$ mm, $L_R = 3.25$ nH, $C_L = 4.1$ pF, $C_R = 1.3$ pF, and $L_L = 5.2$ nH. These parameters were extracted from the unit cell's simulation model using HFSS™. The prototype microstrip antenna, shown in **Figure 1**(c), was constructed using standard photolithographic techniques. Each of the unit cell has an area of 2.5 × 5.2 mm^2 ($0.003\lambda_0 \times 0.006\lambda_0$, where λ_0 is free-space wavelength at 400 MHz). Considering the size of SMD1206 (3.4 mm), the overall size of the antenna is 13.4 × 5.2 × 1.6 mm^3 or $0.017\lambda_0 \times 0.006\lambda_0 \times 0.002\lambda_0$.

It is well established that antennas that operate using a resonance configuration have a limited bandwidth that results from destructive interference of the waves. However, by blocking the standing waves and instead using traveling waves can enhance the bandwidth of the antenna as shown below. As mentioned earlier, the proposed antenna provides a guiding structure for traveling waves as the main radiating mechanism. In this configuration, the voltage and current are in phase and have the same $e^{-j\gamma z}$ distribution along the length, where $\gamma = Kz = \beta - j\alpha$. Lower limit of the leaky-wave bandwidth is the frequency at which $\alpha = \beta$ and the upper limit is reached when $\beta = K_o$. In this case, the surface currents that generate the RF signal travel through the antenna in one direction from the input port#1 to the termination port#2, which is in contrast to conventional standing wave or resonant antenna, such as the monopole or dipole. The results presented below confirm that the proposed traveling-wave antenna exhibits a wider operational bandwidth.

Figure 3 shows the simulated and measured reflection-coefficient performance of the traveling-wave antenna. The results reveal that the antenna is particularly sensitive at the following frequencies: 1.4, 2.65, and 3.7 GHz. Simulation results show the antenna has an operating bandwidth of 4.55 GHz from 300 MHz to 4.85 GHz for $S_{11} < -10$ dB. The equivalent fractional bandwidth of the antenna is 176.7%. The measured results show the antenna operates over 400 MHz to 4.7 GHz (for $S_{11} < -10$ dB), which corresponds to a fractional bandwidth of 168.62%. The discrepancy in the simulation and measured results is attributed to manufacturing tolerance and imperfect soldering of the SMA connector.

The measured gain and radiation efficiency of the antenna at various frequencies are tabulated in **Table 1**. The antenna exhibits a maximum gain and radiation efficiency of 2 dBi and 65%, respectively, at 2.5 GHz. The radiation patterns measured in the E-plane and H-plane at 0.4, 1.5, 2.5, and 4.7 GHz are shown in **Figure 4**. These results show that the proposed antenna radiates unidirectionally. It is worth noting that the metamaterial structure which constitutes the antenna has effectively increased the antenna's aperture without increasing its footprint to enhance its gain and radiation characteristics.

http://dx.doi.org/10.5772/intechopen.68600

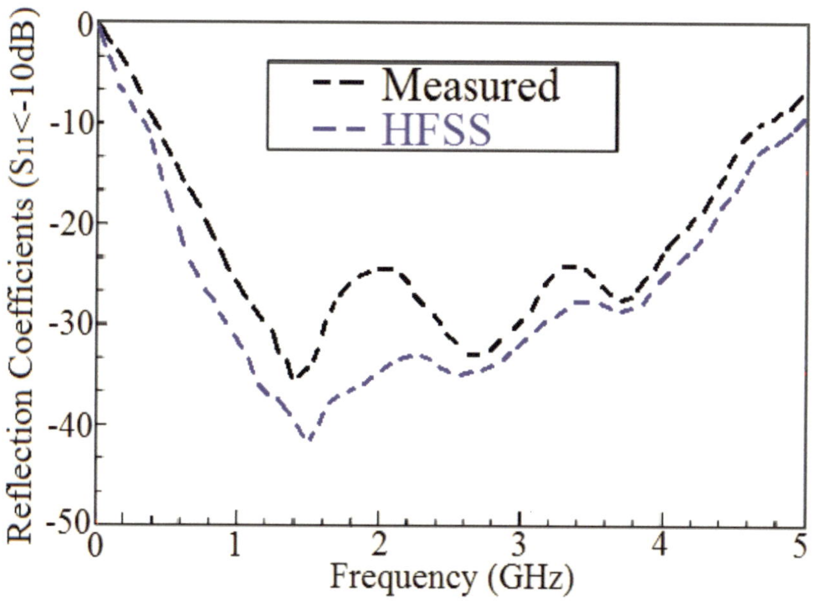

Figure 3. Simulated and measured reflection coefficients response.

Frequency (GHz)	**0.4**	**1.5**	**2.5**	**3.5**	**4.7**
Gain (dBi)	0.05	1.5	2	1.5	0.8
Efficiency (%)	8	45	65	42	30

Table 1. Measured gain and radiation efficiency.

This section presented an innovative metamaterial unit cell. The unit cell is easy to implement by simply etching a slot in a patch radiator, which is short-circuited using a high impedance line to ground through a via-hole. The proposed antenna is a nonresonant traveling-wave radiator that is significantly smaller than a conventional traveling-wave antenna. This is because the metamaterial antenna's dimensions are independent of frequency. The advantages of the proposed antenna are: (i) compactness with a footprint of 13.40 × 5.20 × 1.60 mm^3 ($0.017\lambda_0 \times 0.006\lambda_0 \times 0.002\lambda_0$ at 400 MHz); (ii) low-profile structure; (iii) light weight; (iv) wideband operation from 400 MHz to 4.7 GHz, which corresponds to a large fractional bandwidth of ~170%; (v) unidirectional radiation patterns in both E-plane and H-plane over its operational frequency bandwidth; and (vi) maximum gain and radiation efficiency of 2 dBi and 65%, respectively, at 2.5 GHz. These characteristics make the antenna attractive for use in multiple wireless communication systems.

The proposed traveling-wave antenna's measured performance is compared with other metamaterial and conventional antennas in **Table 2**. It is evident from these results the antenna's operating frequency and fractional bandwidth are superior to metamaterial antennas reported

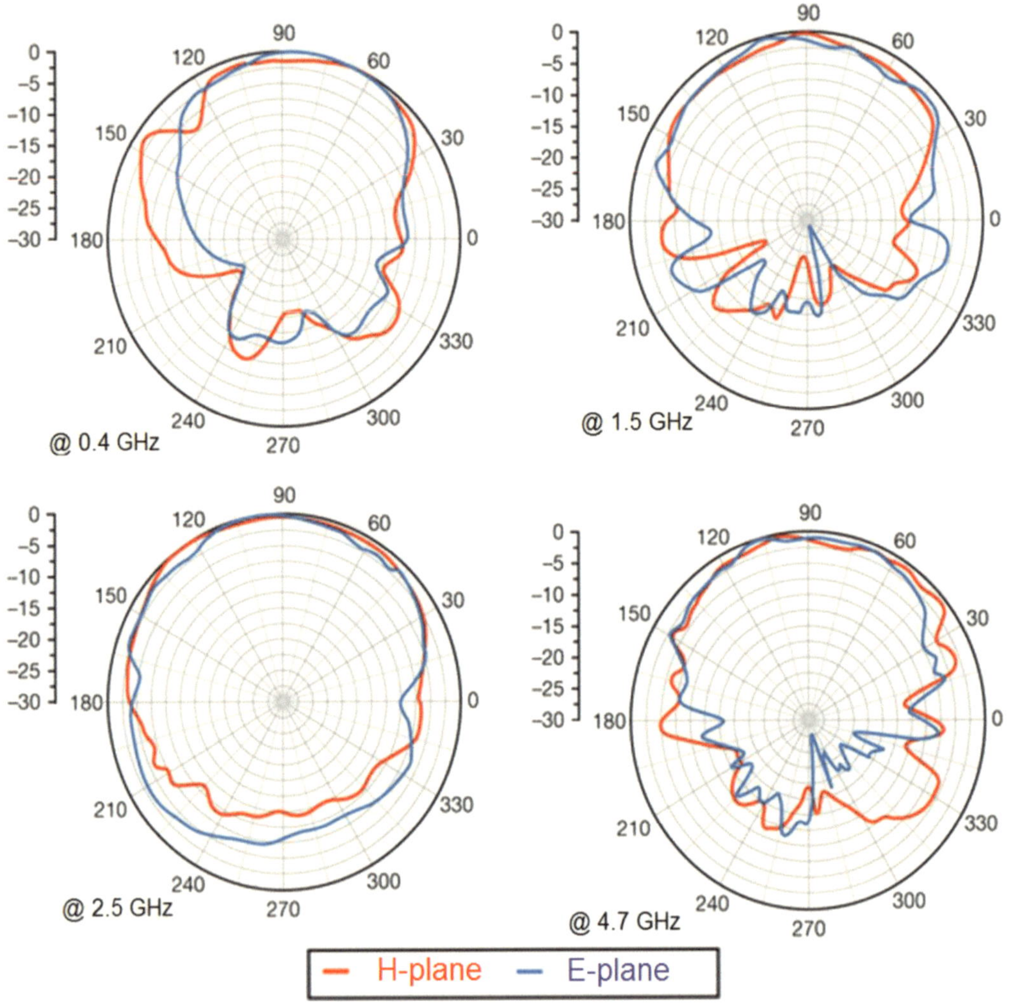

Figure 4. Measured E-plane and H-plane radiation patterns of the proposed antenna.

in Refs. [1, 2]. Its operating frequency band is better than antennas reported in Refs. [1, 2] by a factor of 2.87 and 2.15, respectively; and the fractional bandwidth is better than antennas presented in Refs. [1, 2] by a factor of 7.33 and 4.8, respectively.

To summarize, a novel traveling-wave planar printed antenna is presented in this section based on metamaterial technology. The antenna is shown to provide wideband performance with enhanced gain and radiation efficiency. The low-profile and miniature antenna consists of coupled metamaterial unit cells comprising X-shaped slots and shunted inductive stubs, which are implemented on a rectangular microstrip patch. The proposed antenna is easy to fabricate using conventional PCB manufacturing techniques.

Ref.	Dimensions	Freq. BW/fractional BW	Gain (dBi)	Efficiency (%)
[1]	20.4 × 6.8 × 0.8 mm^3 ($0.39\lambda_0 \times 0.13\lambda_0 \times 0.015\lambda_0$ @ 5.8 GHz)	5.8–7.3 GHz/23%	4.8	78
[2]	7.2 × 5 × 0.8 mm^3 ($0.11\lambda_0 \times 0.079\lambda_0 \times 0.012\lambda_0$ @ 4.7 GHz)	4.7–6.7 GHz/35.08%	3.6	60.3
[11]	12 × 12 × 3.33 mm^3 ($0.09\lambda_0 \times 0.09\lambda_0 \times 0.02\lambda_0$ @ 2.34 GHz)	2.34–2.54 GHz/8.19%	1	20
[12]	20 × 25 × 0.8 mm^3 ($0.22\lambda_0 \times 0.28\lambda_0 \times 0.009\lambda_0$ @ 3.45 GHz)	3.45–3.75 GHz/8.33%	2	25
[13]	60 × 5 × 5 mm^3 ($0.16\lambda_0 \times 0.013\lambda_0 \times 0.013\lambda_0$ @ 0.82 GHz)	0.82–2.48 GHz/100.6%	0.45	53.6
This work	13.4 × 5.2 × 1.6 mm^3 ($0.017\lambda_0 \times 0.006\lambda_0 \times 0.002\lambda_0$ @ 0.4 GHz)	0.4–4.7 GHz/168.62%	2	65

Table 2. Comparison of the metamaterial traveling-wave antenna with other planar antennas (gain and efficiency are optimum values).

3. Slotted patch antenna

Slotted patch antennas are excellent candidates for broadband applications [9, 14–24]. In this section, a compact planar slotted antenna (PSA) is designed and characterized. The antenna is developed using unique metamaterial structures that can be realized using distributed transmission lines constituted from unit cells consisting of series capacitor and a shunt inductor [1]. As mentioned earlier, in reality, the loss associated with metamaterial structures introduces series parasitic inductance in the capacitor and parasitic shunt capacitance in the inductor. The resulting structure is essentially a composite of right- and left-handed (CRLH) transmission lines [25, 26].

In this chapter, the metamaterial unit cell is realized directly on a patch antenna using a combination of capacitive slot etched on the patch and a short-circuited inductive stub that is realized using a high impedance spiral-shaped transmission line. The proposed antenna is modeled and analyzed using ANSYS HFSS™. The antenna's impedance bandwidth, gain, and radiation efficiency were optimized using HFSS™. In fact, the analysis showed that the antenna's performance can be achieved by simply increasing the number of inductive stubs, which does not affect the antenna's overall size. This is because the metamaterial unit cells in essence increase the aperture of the antenna. The optimized antenna is shown to operate over 0.75–4.5 GHz with a peak gain of 3.5 dBi and radiation efficiency of 60% at 2 GHz. The resulting antenna is highly compact and is fairly easy to manufacture.

The proposed antenna had to operate in the UHF and SHF frequency bands with good radiation characteristics, and had to fit inside an area of 25 × 10 mm^2. In order to realize these stringent requirements, the antenna was designed using metamaterial unit cells consisting of series left-handed (LH) capacitors and shunt left-handed inductors. The left-handed capacitors were realized by etching slots on the patch, and the left-handed inductors were realized with a short-circuited high impedance line connected to the patch.

Two antenna prototype designs were evaluated. The first prototype antenna consisted of five unit cells etched on the rectangular patch. Configuration of each unit cell comprised symmetrical arrangement of slot—grounded inductor, where the shape of the slot resembles a crochet note "d," and the inductor is a coiled stub that is grounded through a via-hole. Simulation analysis showed the unusual shape of the slot contributed in enhancing the antenna's performance. **Figure 5** shows the fabricated planar slotted antennas and their equivalent electrical circuit models. Surface currents flowing over the antenna develop a voltage gradient between the metallization and the ground plane to induce parasitic right-handed reactance components, which are represented by the series inductance (L_R) and the shunt capacitance (C_R). Right-handed loss in the structure is represented by R_R and G_R, and left-handed loss by components G_L and R_L. Also included are dielectric loss associated with C_L and the ohmic loss associated with L_L. The planar slotted antenna was implemented on Rogers RO4003® dielectric substrate with permittivity of 3.38, height of 1.60 mm and tanδ of 0.0022. The values of the parameters depicted in **Figure 1** were obtained from HFSS™ analysis, which are: $C_L = 5.10$ pF, $L_L = 6.50$ nH, $C_R = 2.50$ pF, $L_R = 4.90$ nH, $G_L = 7$ S, $G_R = 4.50$ S, $R_L = 7.50\ \Omega$, and $R_R = 5.50\ \Omega$.

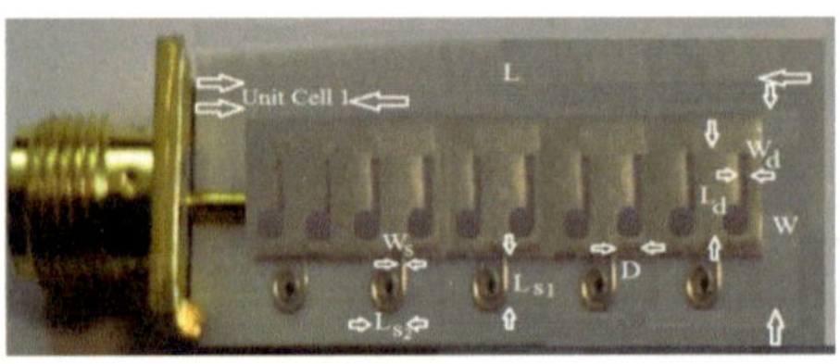

Manufactured prototype of Ant.#1

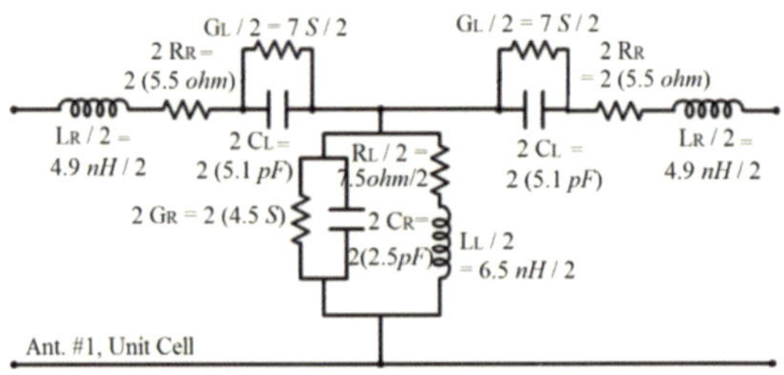

Equivalent circuit of the unit cell in Ant.#1

(a)

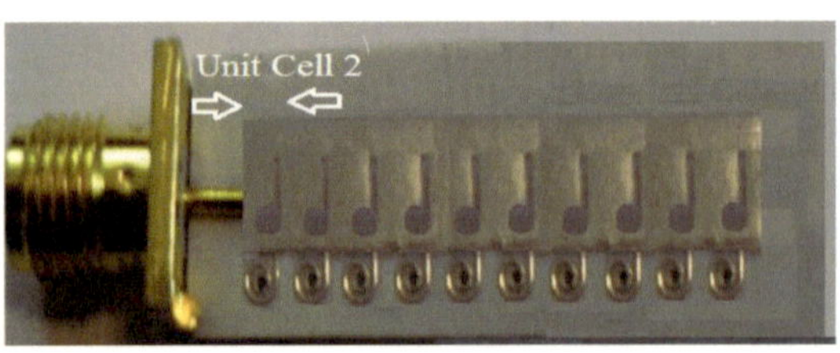

Manufactured prototype of Ant.#2

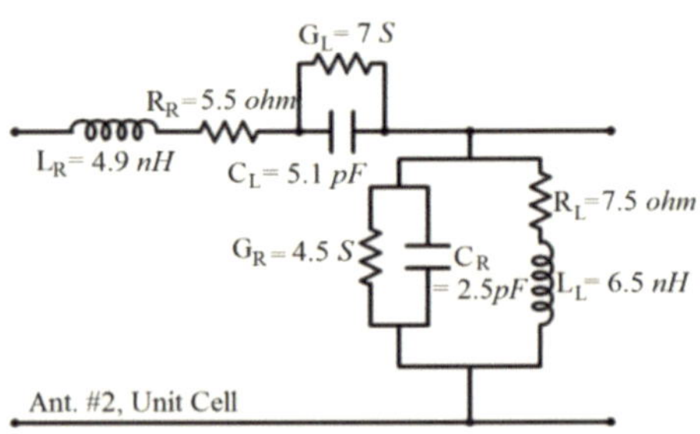

Equivalent circuit of the unit cell in Ant.#2

(b)

Figure 5. The two planar slotted antenna configurations. (a) Ant.#1 based on five symmetrical unit cells, (b) Ant.#2 based on 10 asymmetrical unit cells. The size of this antenna is the same as Ant.#1.

http://dx.doi.org/10.5772/intechopen.68600

Excitation of the antenna through an SMA connector can cause current imbalance resulting in currents flowing over the connector leading to spurious radiation that would degrade the radiation characteristics of the antenna. To prevent this from happening, a second antenna was designed using asymmetrical metamaterial unit cells consisting of slot-grounded inductor configuration, as shown in **Figure 5**(b). The unit cell is constituted from an arrangement of series left-handed capacitor and shunt left-handed inductance. The equivalent electrical model includes loss components represented by R_R, G_R, G_L, R_L, C_L, and L_L. The inductive coils contain the electromagnetic energy near antenna structure and prevent unwanted coupling from happening between the antenna and SMA connector, which helps to enhance the radiation characteristics of the antenna. The antenna's physical and equivalent circuit parameters are given in **Table 3**. The dimensions of antenna and its ground plane size are 25 × 10 × 1.60 mm^3 and 28.50 × 14.50 mm^2, respectively.

The simulated and measured reflection coefficient performance of the prototype antennas are shown in **Figure 6**. The bandwidth of Ant.#1 is 3.2 GHz for $S_{11} < -10$ dB; however, Ant.#2 provides a wider bandwidth of 3.75 GHz. The gain and radiation efficiency of both antennas in **Figure 7** shows that at 2 GHz, Ant.#1 has a maximum gain and radiation efficiency of 1.5 dBi and 35%, respectively; whereas Ant.#2 performs much better with a maximum gain and radiation efficiency of 3.5 dBi and 60%, respectively. These results show that the antenna's aperture can be increased by simply increasing the number of inductive stubs in the proposed antenna. The discrepancy between the simulated and measured performance in **Figure 6** is prominent at lower frequencies. This is attributed to the feed cable and spurious electromagnetic coupling

UC1	UC2	L_g	W_g	0	L_d	L_{S1}	L_{S2}	W	W_d
5	10	28.50	14.50	25	4	3	1.7	10	0.40
Ws	*D*	C_L	L_L	C_R	L_R	G_L	G_R	R_L	R_R
0.20	1.40	5.10	6.50	2.50	4.90	7	4.50	7.50	5.50

Table 3. Design parameters (units in mm, pF, nH, S, Ω; unit cell is abbreviated as UC).

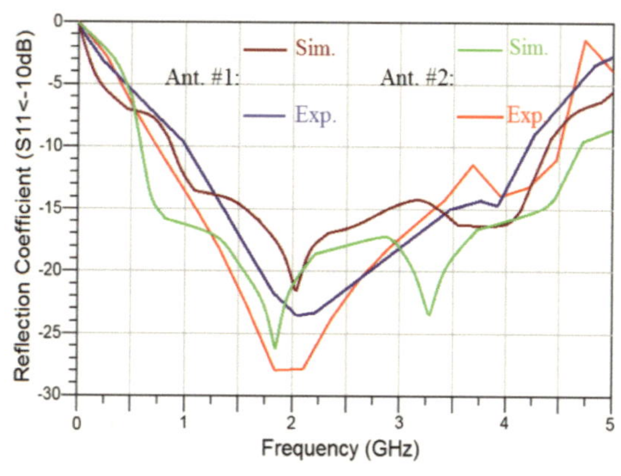

Figure 6. Simulated and measured reflection coefficient response of the two prototype antennas.

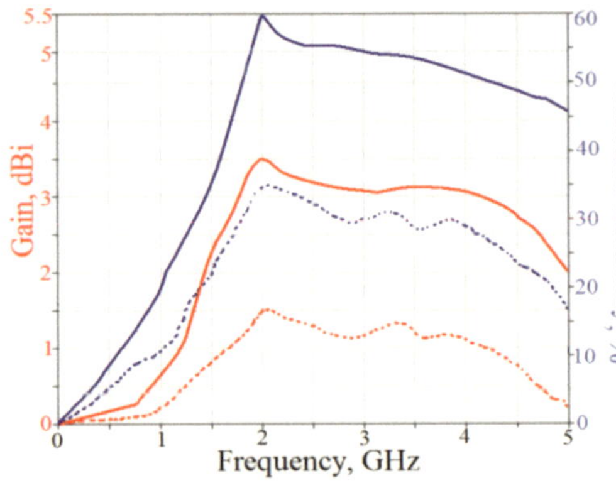

Figure 7. Measured gain and radiation efficiency response of the two prototype antennas. (Ant.#1: dashed lines, and Ant.#2: solid lines).

with the SMA connector. In the simulation, the antenna is excited with a signal source without using a feeding cable. However, in practice, the measurement of the prototype antennas was measured with Agilent's 8722ES vector network analyzer connected to the antenna using a coaxial cable. The salient results are given in **Table 4**.

The measured radiation patterns of Ant.#1 and Ant.#2 in the E-plane and H-plane are shown in **Figure 8**. The radiation pattern of Ant.#1 is predominately bidirectional. At 1 GHz, the E-plane and H-plane radiation patterns of Ant.#1 are analogous to a dipole antenna with a maximum gain at an angle of 180 degrees azimuth; however, at 2 GHz, the maxima gain flips to 0 degrees. At 4 GHz, the maxima gain flips back to at 180 degrees. This phenomenon is observed because of the backward and forward wave variation of the phase with frequency. Ant.#2 exhibits bidirectional radiation patterns in the E-plane and H-plane at 1 GHz; however, the radiation pattern changes to omnidirectional at 2 GHz. The additional inductive stubs essentially suppress the effects of current imbalance over the SMA connector, and hence, improve its impedance bandwidth. At 2 GHz, the radiation gain of Ant.#2 in the H-plane is reduced by about 3 dB at 90 and 270 degrees; and at 4 GHz, a null is observed in the E-plane at 180 degree.

	Ant.#1	Ant.#2	Change
Dimensions	25 × 10 × 1.60 mm^3 or $0.083\lambda_0 \times 0.033\lambda_0 \times 0.005\lambda_0$ @ 1 GHz Ground-plane of ants.: 28.50 × 14.50 mm^2 or $0.098\lambda_0 \times 0.048\lambda_0$		None
Bandwidth	Sim.: 0.89–4.33 GHz Fractional BW ≈ 131.8%	Sim.: 0.58–4.7 GHz Fractional BW ≈156.06%	24.26%
	Meas.: 1–4.20 GHz Fractional BW ≈ 123.07%	Meas.: 0.75–4.50 GHz Fractional BW ≈142.85%	19.78%
Gain (dBi)	0.20, 1.50, and 1	0.25, 3.50, and 2.70	2 dB increase @ 2 GHz
Efficiency (%)	11, 35, and 27	13, 60, and 48	25% increase @ 2 GHz

Note: Gain and radiation efficiency were measured at 1, 2, and 4.2 GHz for ant.#1; and at 0.75, 2, and 4.5 GHz for ant.#2.

Table 4. Salient features of the proposed planar slotted antenna.

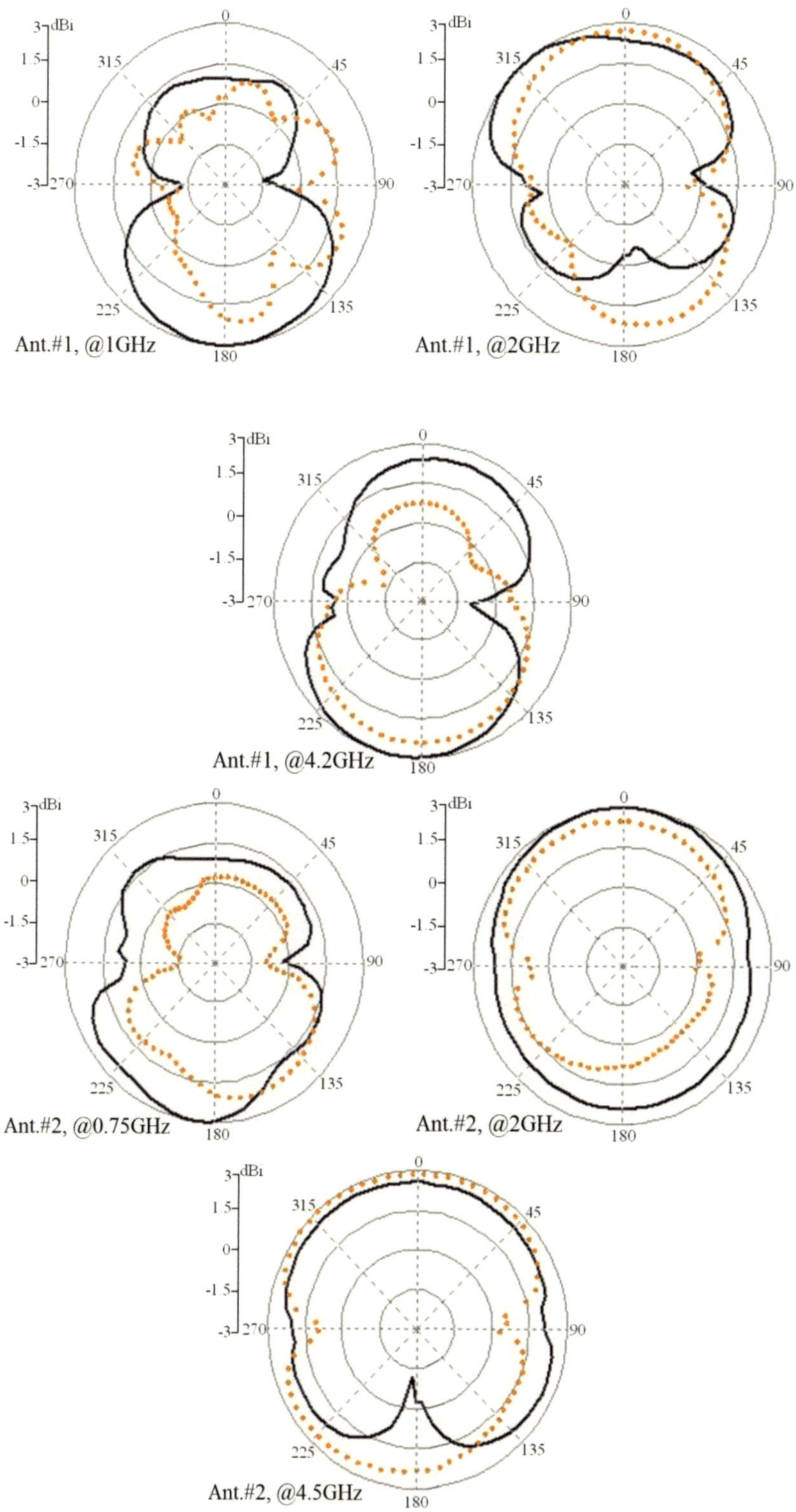

Figure 8. Measured E-plane and H-plane radiation patterns of Ant.#1 and Ant.#2 at various frequencies.

To summarize, novel planar slotted antenna designs were presented based on metamaterial unit cells that were constituted from capacitive slots etched in the radiating patch and grounded coiled inductive stubs. The fabricated antennas possess the following characteristics: (i) compact structure; (ii) simple layout, which is easy to fabricate at low cost; (iii) low-profile structure that is easy to integrate with RF circuits; and (iv) characteristics suitable for UHF-SHF broadband wireless communication systems.

Author details

Mohammad Alibakhshikenari[1]*, Mohammad Naser-Moghadasi[2], Ramazan Ali Sadeghzadeh[3], Bal Singh Virdee[4] and Ernesto Limiti[1]

*Address all correspondence to: alibakhshikenari@ing.uniroma2.it

1 Department of Electronic Engineering, University of Rome "Tor Vergata", Rome, Italy

2 Faculty of Engineering, Science and Research Branch, Islamic Azad University, Tehran, Iran

3 Faculty of Electrical Engineering, K. N. Toosi University of Technology, Tehran, Iran

4 London Metropolitan University, Center for Communications Technology & Mathematics, London, UK

References

[1] Alibakhshi-Kenari M, Naser-Moghadasi M, Virdee BS, Andújar A, Anguera J. Compact antenna based on a composite right/left handed transmission line. Microwave and Optical Technology Letters. 2015; **57**(8):1785-1788

[2] Alibakhshi-Kenari M. New traveling-wave antenna resonating at 6 GHz based on artificial transmission line metamaterial structures for RF portable devices. Open Journal of Antennas and Propagation. 2013;**1**(2). DOI: 10.4236/ojapr.2013.12003

[3] Collin RE. Field Theory of Guided Waves. 2nd ed. United States: Wiley-Interscience; 1991. chap. 12

[4] Balanis CA. Antenna Theory and Design. United States: John Wiley & Sons; 1997

[5] Pozar DM. Microwave and RF Design of Wireless Systems. New York: John Wiley & Sons, Inc. 2000

[6] Pozar DM. An update on microstrip antenna theory and design including some novel feeding techniques. IEEE Antennas and Propagation Society Newsletter. 1986; **28**(5): 4-9

[7] Bahl IJ, Bhartia P. Microstrip antennas. United States: Artech House; 1980

[8] Alibakhshi-Kenari M, Naser-Moghadasi M, Sadeghzadah RA. Bandwidth and radiation specifications enhancement of monopole antennas loaded with split ring resonators. IET Microwaves, Antennas & Propagation. 2015;**9**(14):1487-1496

[9] Engheta N, Ziolkowski RW editors . Electromagnetic Metamaterials: Physics and Engineering Explorations. Hoboken: Wiley and IEEE Press; 2006

[10] Caloz C, Itoh T. Electromagnetic Metamaterials, Transmission Line Theory and Microwave Applications. Wiley and IEEE Press; 2005

[11] Lee CJ, Leong KM H, Itoh T. Broadband small antenna for portable wireless application. In: International Workshop on Antenna Technology: Small Antennas and Novel Metamaterials; 4-6 March 2008; Chiba, Japan: IEEE; 2008. 10-13

[12] Yu C-C, Huang M-H, Lin L-K, Chang Y-T. A compact antenna based on MTM for WiMAX. In: Asia-Pacific Microwave Conference; 16-20 December 2008; Macau, China: IEEE; 2008.pp. 1-4

[13] Li Y, Zhang Z, Zheng J, Feng Z. Compact heptaband reconfigurable loop antenna for mobile handset. IEEE Antennas and Wireless Propagation Letters. 2011;**10**:1162-1165

[14] Liu L, Cheung SW, Yuk TI. Bandwidth improvements using ground slots for compact UWB microstrip-fed antennas. Progress in Electromagnetics Research Symposium. 2011;1420-1423

[15] Cheung SW, Liu L, Azim R, Islam MT. A compact circular-ring antenna for ultra-wideband applications. Microwave and Optical Technology Letters. 2011;**53**:2283-2288

[16] Sun YY, Cheung SW, Yuk TI. Studies of Planar Antennas with different radiator shapes for ultra-wideband body-centric wireless communications. Progress in Electromagnetics Research Symposium. 2011:1415-1419

[17] Zhang J, Sun XL, Cheung SW, Yuk TI. CPW-coupled-fed elliptical monopole antenna for UWB applications. IEEE Radio Wireless Week. 2012;295-298

[18] Sun YY, Islam MT, Cheung S W, Yuk TI, Azim T, Misran N. Offset-fed UWB antenna with multi-slotted ground plane. IEEE International Workshop on Antenna Technology. 2011:432-436

[19] Sun YY, Cheung SW, Yuk TI., Planar monopole ultra-wideband antennas with different radiator shapes for body-centric wireless networks. Progress in Electromagnetics Research Symposium. 2012:839-843

[20] Wong H, So KK, Gao X. Bandwidth enhancement of a monopolar patch antenna with V-shaped slot for car-to-car and WLAN communications. IEEE Transactions on Vehicular Technology. 2016;**65**(3):1130-1136

[21] Liu S, Wu W, Fang DG. Single-feed dual-layer dual-band E-shaped and U-slot patch antenna for wireless communication application. IEEE Antennas and Wireless Propagation Letters. 2016;**15**:468-471

[22] Chen J, Tong K-F, Al-Armaghany A, Wang J. A dual-band dual-polarization slot patch antenna for GPS and Wi-Fi applications. IEEE Antennas and Wireless Propagation Letters. 2016;**15**:406-409

[23] Cao W-Q, Hong W, Cai Y. Microstrip-line-slot fed EDPA elements and array with ELP. IET Microwaves, Antennas & Propagation. 2016; **10**(12):1304-1311

[24] Soltanpour M, Fakharian MM. Compact filtering slot antenna with frequency agility for Wi-Fi/LTE mobile applications. Electronics Letters. 2016;**52**(7):491-492

[25] Eleftheriades GV. EM transmission-line metamaterials. Materials Today. 2009;**12**(3):30-41

[26] Lai A, Caloz C, Itoh T. Composite right/left handed transmission line metamaterials. IEEE Microwave Magazine. 2004;**5**(3):34-50

Chapter 7

Low-SAR Miniaturized Handset Antenna Using EBG

Kamel Salah Sultan, Haythem Hussien Abdullah and
Esmat Abdel-Fatah Abdallah

Additional information is available at the end of the chapter

http://dx.doi.org/10.5772/intechopen.70175

Abstract

Advances in wireless communications have paved the way for wide usage of mobile phones in modern society, resulting in mounting concerns surrounding its harmful radiation. Energy absorption in human biological tissues can be characterized by specific absorption rate (SAR). This value refers to the actual amount of electromagnetic energy absorbed in the biological tissues, thus a lower value of SAR indicates a lower radiation exposure risk to the human body. So, our challenge is to introduce mobile handset antennas with low SAR and operating at all mobile and wireless applications. In this chapter, novel configurations of single-element antenna are designed, simulated, fabricated, and measured. The antennas operate for most cellular applications: global system for mobile (GSM)-850/900, digital cellular system (DCS)-1800, personal communication service (PCS)-1900, universal mobile telecommunication system (UMTS)-2100, and long-term evolution (LTE) bands. The antennas also support wireless applications. The proposed antennas have a compact size and low SAR at all bands. Also, this chapter presents a comprehensive study on the performance of the antenna in the different environments. Furthermore, the antenna performance is tested in the presence of head and hand in free space and in a car. The simulation and measurement results are in good agreement.

Keywords: specific absorption rate (SAR), multiband antenna, monopole, meander, long-term evolution (LTE), industrial, scientific and medical (ISM), wireless local area network (WLAN), electromagnetic bandgap (EBG), printed antenna

1. Introduction

With the rapid growth of communication technologies and the vast increase of mobile services, developing new low specific absorption rate (SAR) antenna with compact size becomes of great demand in the international market. Furthermore, the new personal mobile handsets are needed to support multimedia (image with good resolution, clear voice, and data communication) anytime anywhere. This announces that new mobile devices are required to back various technologies and to operate in various bands. So, the long-term evolution (LTE) is presented as new mobile generation to give high performance for communication systems. It has high capacity and large speed of mobile networks [1–6].

Vast research works are introduced to diminish the handset antenna in size and cost and to increase the bands that are provided by the antennas. According to Ang et al. [7], an antenna is introduced to cover four bands (global system for mobile (GSM)-900, digital cellular system (DCS)-1800, personal communication service (PCS)-1900, and industrial, scientific and medical (ISM)-2450) at −6 dB as a reference, and this antenna consists of two-layer folded patches with size of 33 × 16 × 8 mm^3. Ciais et al. [8] replace the ISM 2450 band covered in [7] by the UMTS 2100 band at 6-dB bandwidth, the antenna consists of patch with three extra parasitic elements put on the corner of a ground plane. But, the antenna in Ref. [8] has a double size of the antenna in Ref. [7]. Another two different quad-band antennas are introduced in Refs. [9, 10]. According to Tzortzakakis [9], the proposed antenna consists of a monopole and a helix with half size reduction compared to the antenna in Ref. [7]. According to Ku et al. [10], the proposed antenna consists of a folded dual loop antenna with slight size reduction compared to the antenna in Ref. [7]. According to Tang et al. [11], the proposed compact antenna consists of double inverted L-shape, three-meandered strip and a coplanar strip. The antenna covers five bands (GSM-850, GSM-900, DCS-1800, PCS-1900, and the UMTS-2100) (at −6 dB as reference) with 60% reduction in size referred to the antenna in Ref. [9]. According to Li et al. [12], a folded loop antenna is introduced to cover the hepta band (GSM-850, GSM-900, GSM-1800, GSM-1900, UMTS-2100, GPS, and WLAN at −6 dB bandwidth) with slight size increase compared to the antenna in Ref. [11]. According to Young et al. [13], an octa-band antenna with more compact size of 46 × 7 × 11 mm^{3}is introduced. The antenna covers the following bands: LTE-700, GSM-850, GSM-900, DCS-1800, PCS-1900, UMTS-2100, LTE-2300, and LTE-2500. The antenna introduced by Young et al. [13] has a compact size in addition to covering eight bands for different mobile applications, yet still several bands need to be covered. In addition, it is complicated in the fabrication process due to its multilayer construction.

The vast development of wireless services needs to consider the interactions between the human body, especially human head, and mobile handset, while the human head absorbs the electromagnetic wave that is radiated by the antenna. Some mobile handset antenna characteristics are alerted for its closeness to the human head because of their radiation pattern, radiation efficiency, bandwidth, and return loss. The mutual effects of human head and the antenna have been introduced by many researches [4–6, 14–16].

There are different methods that were used through the last few decades to reduce the SAR such as using other elements as auxiliary antenna, loading a new material such as ferrite, using the electromagnetic bandgap (EBG) structures to reduce surface wave/artificial magnetic conductors (AMC) surfaces, and finally using metamaterials [3, 4]. A combination of the main antenna and a director or a reflector was introduced in Ref. [17] to increase the effective radiation efficiency and to reduce the SAR. The distance between the antenna and the auxiliary elements is considered as the main drawback of this method for increasing the size and cost of the antenna. According to Jung and Lee [18], the ferrite sheet was used as a protection wall between the antenna and the human body. The disadvantage of the procedure is the utilization of a costly ferrite material that has unique properties of permittivity and permeability to achieve low SAR [3]. In Refs. [3–6], the EBG and metamaterial techniques are used due to their properties. The EBG technique reduces the SAR values up to 75%.

2. Multiband antennas

2.1. First configuration

Sultan et al. [4] introduce printed multiband antennas to cover the most of mobile bands and wireless application. **Figure 1(a)** shows the planar monopole antenna that is inverted L-shape in size and has an electrical length of a monopole of quarter-wavelength at 2350 MHz. The monopole operating bands are (1700–3000) MHz and (4600–5500) MHz. The dimensions of the monopole antenna are 18 ×6 × 1.5 mm^3. The antenna is designed with compact dimensions of 20 × 33 × 1.5 mm^3. So, the antenna can be easily integrated in small and sleek mobile device. All the labeled dimensions are tabulated in **Table 1**. **Figure 1(b)** shows the full geometry of the mobile with size of 45 × 110 × 1.5 mm^3 in conjunction with a prototype of the antenna. The antenna is fabricated using the FR4 substrate (ε_r= 4.65) with 1.5-mm thickness and loss tangent of 0.025 as shown in **Figure 1(c)**.

The antenna is composed of a planar monopole and a planar meander line. The meander line increases the path over which the surface current flows and that eventually results in

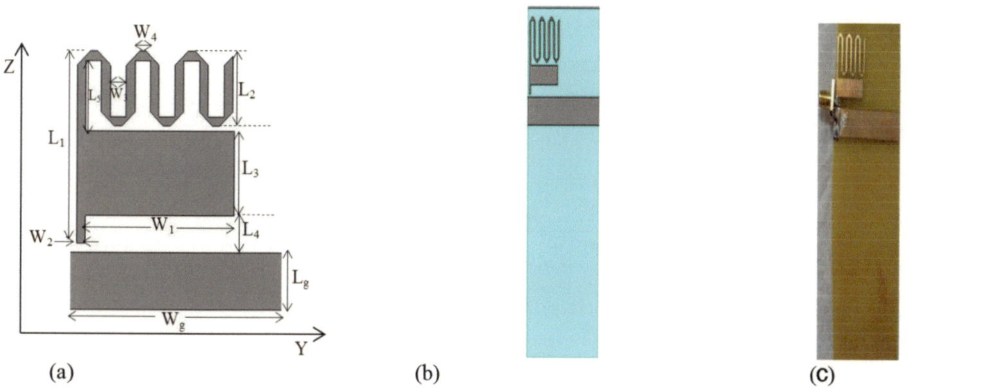

Figure 1. Geometry of the first configuration antenna in mobile phone. (a) Antenna geometry, (b) antenna with full substrate, and (c) antenna prototype.

lowering the resonant frequency. The optimized length of the meander line is 112 mm to operate at 900 MHz. Furthermore, it operates also at the higher frequencies. Where the combination between the monopole and meander line contributes to operate from 1.7 to 7.4 GHz, the distance between the meander line and the monopole is considered as 1 mm that controls the bandwidth matching. Also, the ground plane was chosen to be coplanar with the monopole with size 45 × 9 mm^2.

The simulated and measured return loss results are introduced in **Figure 2**. One can notice that the simulated and the measured results cover all the proposed mobile and wireless application bands. The antenna operates in the two bands: lower band (0.870–0.970 GHz) and higher band (1.6349–7.4233 GHz) when taking the 6 dB return loss as a reference.

2.2. Second and third configurations

One of the disadvantages of the first configuration antenna is the low front to back ratio and its large SAR value as reported in Ref. [3]. So, if the EBG structure is applied to the antenna, it reduces the surface waves and prevents the undesired radiation from the ground plane as introduced in Ref. [19]. So, the radiation toward the human head is reduced and, hence, the SAR values are also reduced. Although the EBG technique has the advantages of lowering cost and the ease of implementation, the EBG structure still needed more development to produce a practical wideband and small size for multiband applications. Before applying the EBG structure, the antenna dimensions were 30 × 33 × 1.5 mm^3. These periodic structures have high electromagnetic surface impedance, which is capable of suppressing the propagation of surface currents and acting as a perfect magnetic conductor in a certain frequency range, so the antenna dimensions are reduced to 20 × 26 × 1.5 mm^3. The antenna can be easily integrated in small and sleek mobile device. In this part, two antenna configurations are introduced. In the second configuration, a planar EBG structure is positioned between the user and the handset antenna, while in the third configuration, the EBG structure is positioned in coplanar with the antenna to increase the gain. The planar EBG structure is periodic square patches with dimensions of 9 × 9 mm^2 and gaps of 1 mm. The whole mobile board of the second configuration and the third configuration is shown in **Figures 3** and **4**. In the second configuration, the EBG is fabricated over the FR4 substrate ($\varepsilon_r = 4.5$) with 0.8 mm thickness and loss tangent of 0.025. In the third configuration, the EBG is fabricated in the top layer of the antenna substrate.

Parameter	Value (mm)	Parameter	Value (mm)	Parameter	Value (mm)
L1	24	L5	1	W3	2
L2	14	Lg	9	W4	1
L3	6	W1	17	Wg	45
L4	3	W2	1		

Table 1. Dimensions of first configuration.

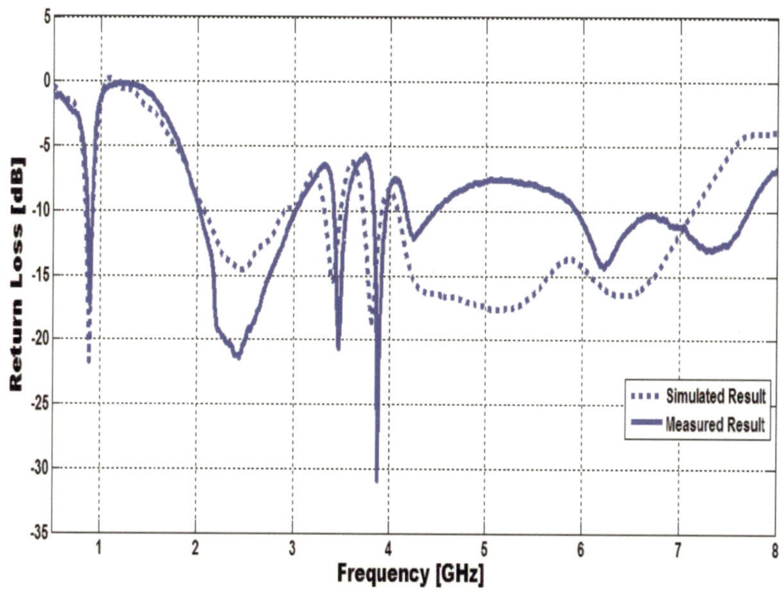

Figure 2. Simulated and measured return loss of the proposed antenna.

The EBG structure is investigated by simulating its behavior within the required band by some iterations. The dimensions of the EBG are selected to be 9 × 9 mm^2 with a separation of 1 mm to achieve a bandgap in the range from 1.54 to 2.48 GHz. The simulated CST and the experimental results of the return loss of the two antennas are shown in **Figure 5**. The results ensure that the antenna covers all the mobile and wireless applications bands. When taking the 6 dB return loss as a reference, the antenna of the second configuration operates in the two bands (850–1030 MHz) and (1.71–7.8 GHz). The antenna of the third configuration operates in the two bands (830–1000 MHz) and (1.67–8.7 GHz). **Figure 6** shows a comparison between the simulated gain of the antenna with and without EBG structures in the two configurations. From the figure, it is noted that the gain is increased with the use of the EBG structure the antennas dimensions are showed in (**Table 2**).

2.3. Fourth configuration

The first configuration antenna was redesigned before applying the EBG on FR4 material with $\varepsilon_r = 4.5$ and $h = 0.8$ mm with compact dimensions of 33 × 25 × 0.8 mm^3. The operating bands are 860–1020 MHz and1.675–8.15 GHz. The EBG structure is embedded on the bottom layer of the substrate as shown in **Figure 7**. The antenna dimensions are showed in **Table 3** for the two cases (with EBG and without EBG). The comparison between the simulated and measured return loss results of the proposed antenna with EBG and without EBG is shown in **Figure 8**. The results reveal that the antenna operates at all the mobile and wireless applications bands. When taking the 6 dB return loss as a reference, the antenna operates in the two bands (from 587 to 977 MHz and from 1.67 to 8.63 GHz). The gain and the radiation efficiency of the fourth antenna are shown in **Table 4**.

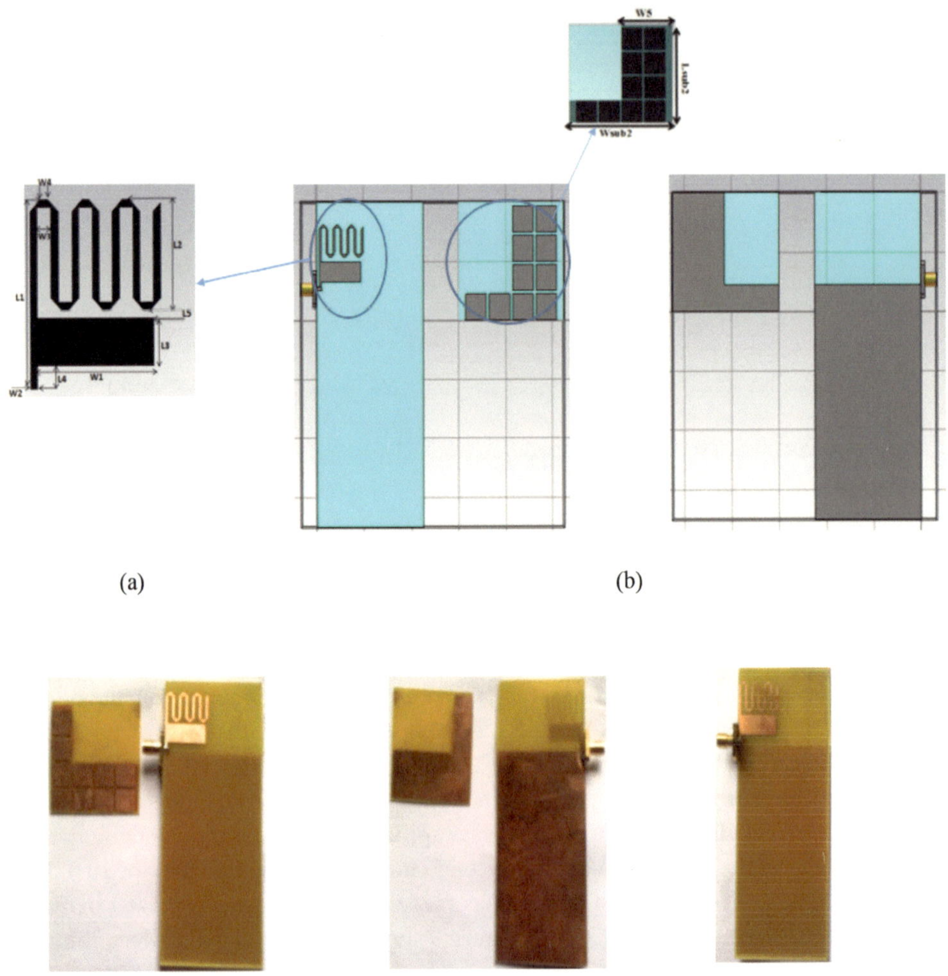

(a) (b)

(c) (d) (e)

Figure 3. Geometry and prototype of second configuration. (a) Front view of second configuration, (b) back view of second configuration (c) front of the two substrates, (d) back of the two substrates, and (e) second configuration assemble.

When EBG structures interact with electromagnetic waves, they show amazing properties such as frequency pass bands, stop bands, or bandgaps. The characteristics of the EBG structure shown in Ref. [4] ensure that the EBG structures have stop bands at most of the mobile applications and wireless applications bands. This ensures that the EBG structures have high surface impedance (HSI) in these bands, and even when a large electric field along the EBG surface is present, the tangential magnetic field is small. Thus, the generated electric field acts as a magnetic current that radiates in conjunction with the original antenna. The EBG structure is positioned perpendicular to the two antennas: the monopole and the meander line. Without the EBG, the monopole and the meander lines have null radiation at the end-fire directions. With the existence of the EBG, the EBG equivalent magnetic current radiates in the end-fire direction of the two antennas. Thus, some of the radiated energy will be in the

http://dx.doi.org/10.5772/intechopen.70175

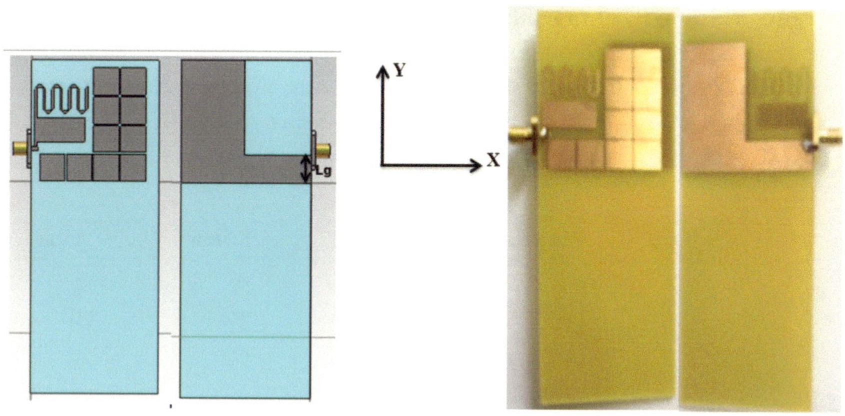

(a) (b)

Figure 4. Geometry and prototype of third configuration. (a) Geometry and (b) prototype.

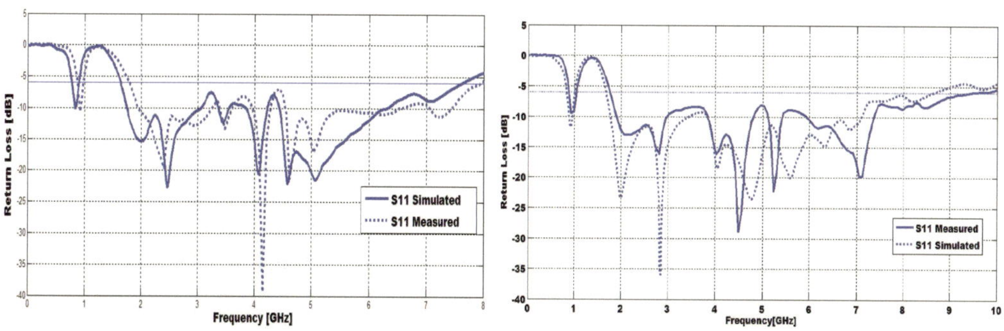

(a) (b)

Figure 5. Simulated and measured return loss of the second and third configuration of proposed antenna. (a) second configuration. (b)Third configuration.

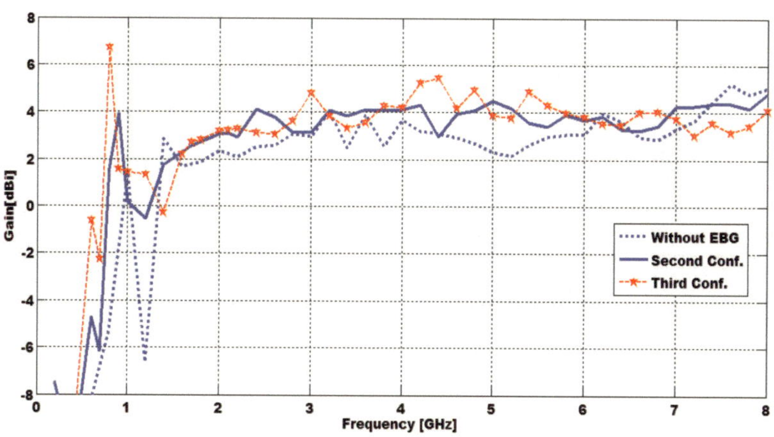

Figure 6. Simulated gain of antennas.

azimuth plane, which reduces the radiation in the elevation direction where the human body exists. Thus, one can conclude that the EBG structure redistributes the radiated energy in such a way that it reduces the radiation toward the human body. **Figure 9** shows the radiation pattern of the proposed antenna in the absence and the presence of the EBG at 1.8 GHz. It is clear that the EBG structures enhance the radiation pattern in the elevation plan. We chose 1.8 GHz

Parameter	Value (mm)	Parameter	Value (mm)	Parameter	Value (mm)
L1	20	L5	1	W3	2
L2	9.5	Lg	79	W4	1
L3	7	W1	17	Wg	45
L4	3	W2	1	Lsub2	40
Wsub2	45	W5	22		

Table 2. Geometric dimensions of the proposed antenna.

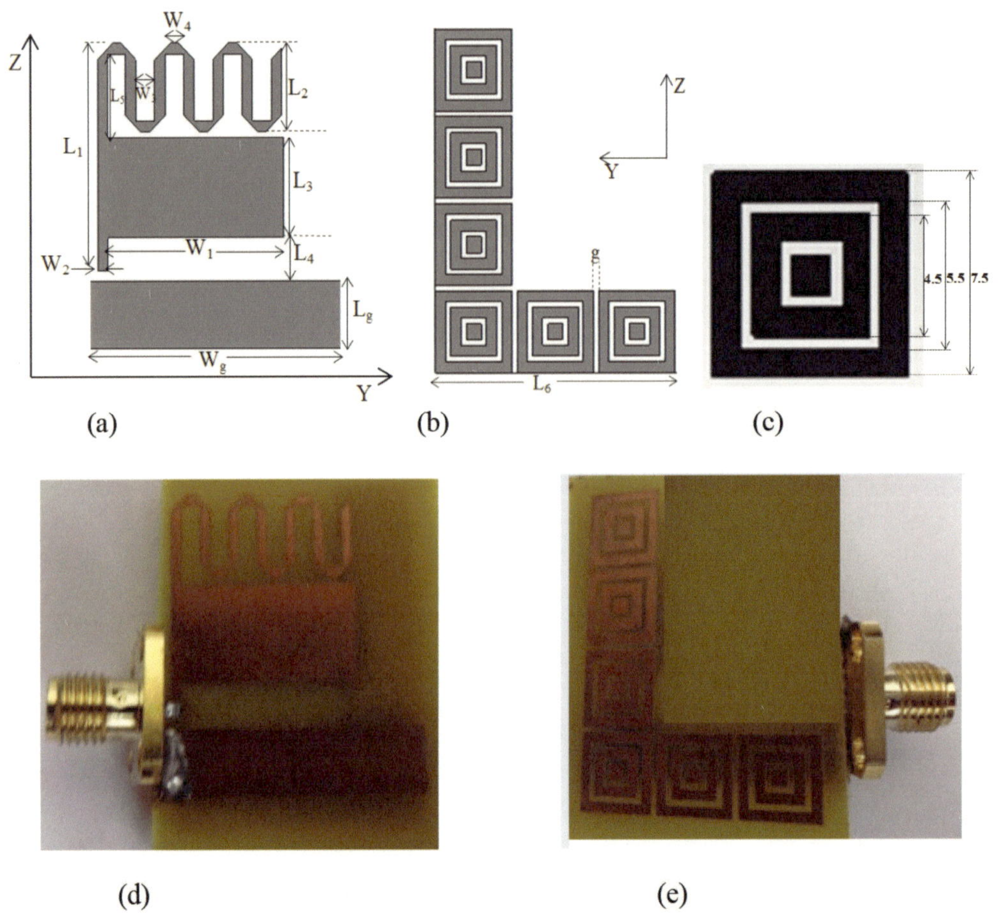

Figure 7. Geometry and prototype of fourth configuration. (a) Front view, (b) back view, (c) unit cell of EBG, (d) front view prototype, and (e) back view prototype.

http://dx.doi.org/10.5772/intechopen.70175

as an example, but the same enhancement is observed for all the mobile application bands. It appears from **Figure 10** that although the radiation pattern is omnidirectional in the elevation plane, it does not have a null at the end-fire direction of angle 90°. **Figure 10** shows the measured and simulated radiation patterns at frequencies 0.9, 1.8, and 2.1 GHz. Radiation pattern measurements were carried out using SATIMO Anechoic antenna chamber.

Parameter	Without EBG	With EBG	Parameter	With EBG
	Value	Value		Value
L_1:L_2	28:14	20.5:8	L_5	7.5
L_3:L_4	6:7	9:3	L_6	23.5
Lg:W_1	6:17	6:17	G	0.5
W_2:W_3	1:2	1:2		
W_4/W_g	1:25	1:25		

Table 3. Antenna dimensions (mm).

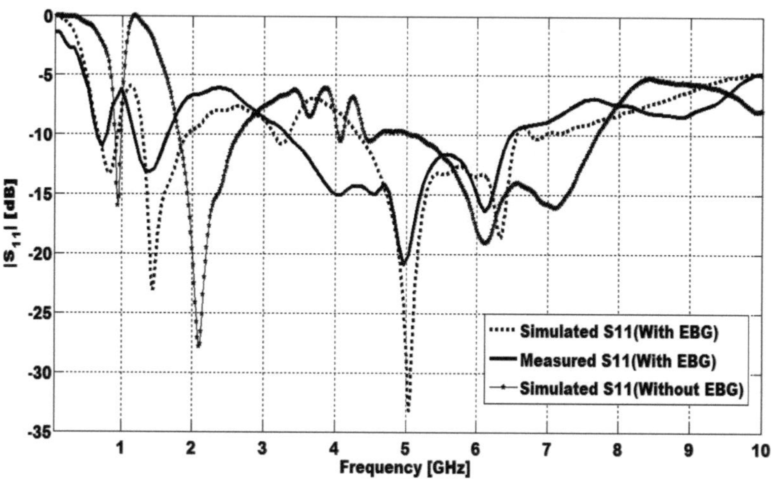

Figure 8. Simulated and measured return loss of the proposed antenna.

F (GHz)		0.9	1.8	2.1
Gain (dBi)	Simulated	2.2	4.4	5.1
	Measured	1.9	4	4.6
Radiation efficiency (%)	Simulated	89	88	81
	Measured	71	68	66

Table 4. Values of gain and radiation efficiency of the proposed antenna.

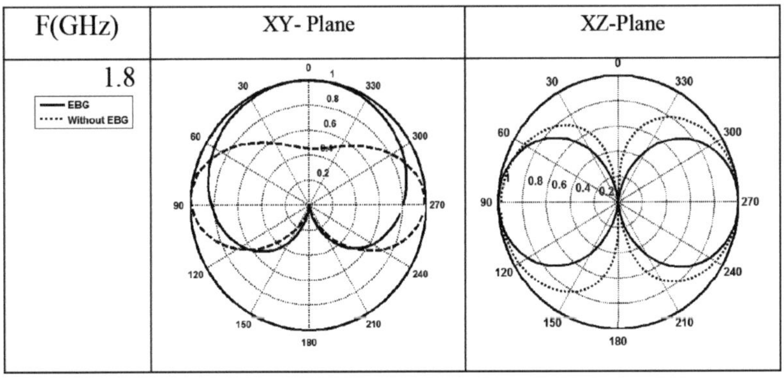

Figure 9. Radiation pattern in the *xy* and *yz* planes. The antenna is in the *yz* plane.

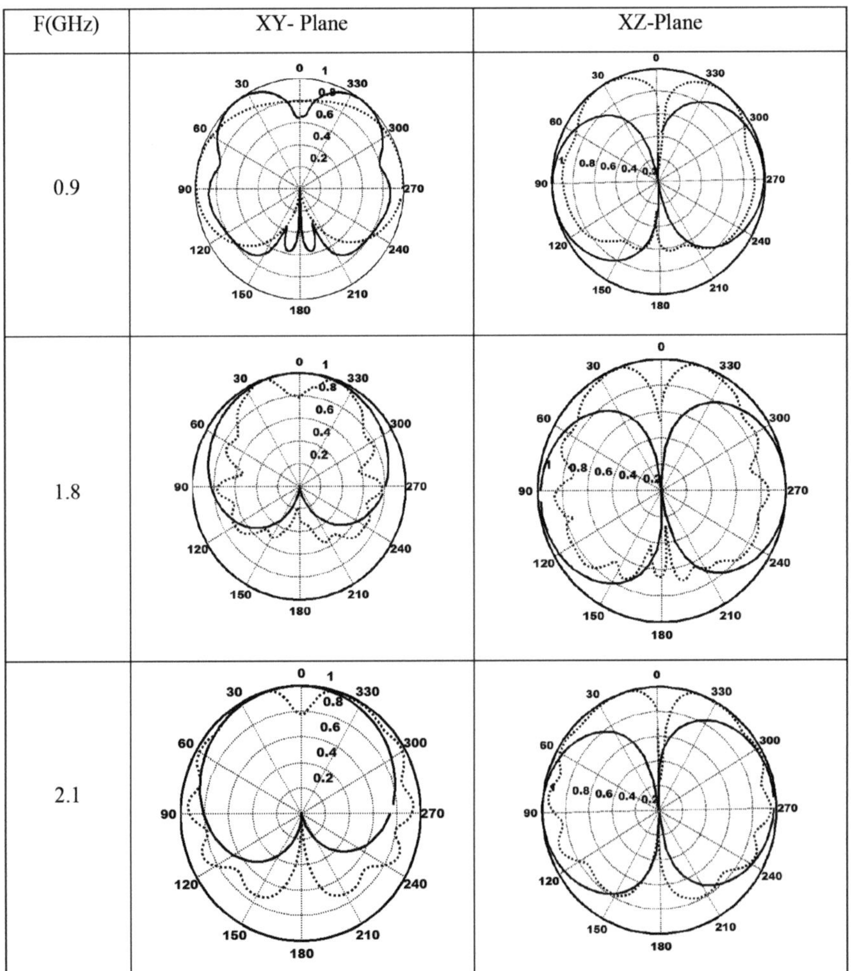

Figure 10. Radiation pattern in the *xy* and *yz* planes (a) 0.9 GHz, (b) 1.8 GHz, and (c) 2.1 GHz. The antenna is in the *yz* plane.

3. SAR calculation

In designing antennas for mobile communications, it is important to investigate the SAR value produced by the radiation from the mobile handsets. The output power of the cellular phone depends on more than one factor such as distance to base station, surrounding environment, frequency band, and technology as shown in Ref. [1]. The reference power of the cellular phone is set to 500 mW at the operating frequencies of 0.9, 1.8, and 2.1 GHz. The SAR values are calculated according to the 10 g standard of the human tissue mass. The SAR calculations are done using the CST Microwave Studio commercial package with Hugo voxel model [20]; the permitivities and the conductivities of the Hugo model tissues are according to the published data in Ref. [21]. The dispersive properties of the tissues are taken into our considerations. As expected, the SAR values depend on many factors such as the operating frequency, the type of antenna, antenna position related to human body, and the relative distance between the human body and the antenna. **Figure 11** shows the SAR calculations on human head model in the presence of the antenna in the *ZY* plane with and without the EBG structure at 0.7, 0.9, 1.8, and 2.1 GHz. As expected, the SAR values depend on the operating frequency, the antenna types, and the distance between the antenna and the human body. **Table 5** shows the averaged 10 g SAR at the aforementioned operating frequencies when the antenna is close to the body. As the results are scalable when the

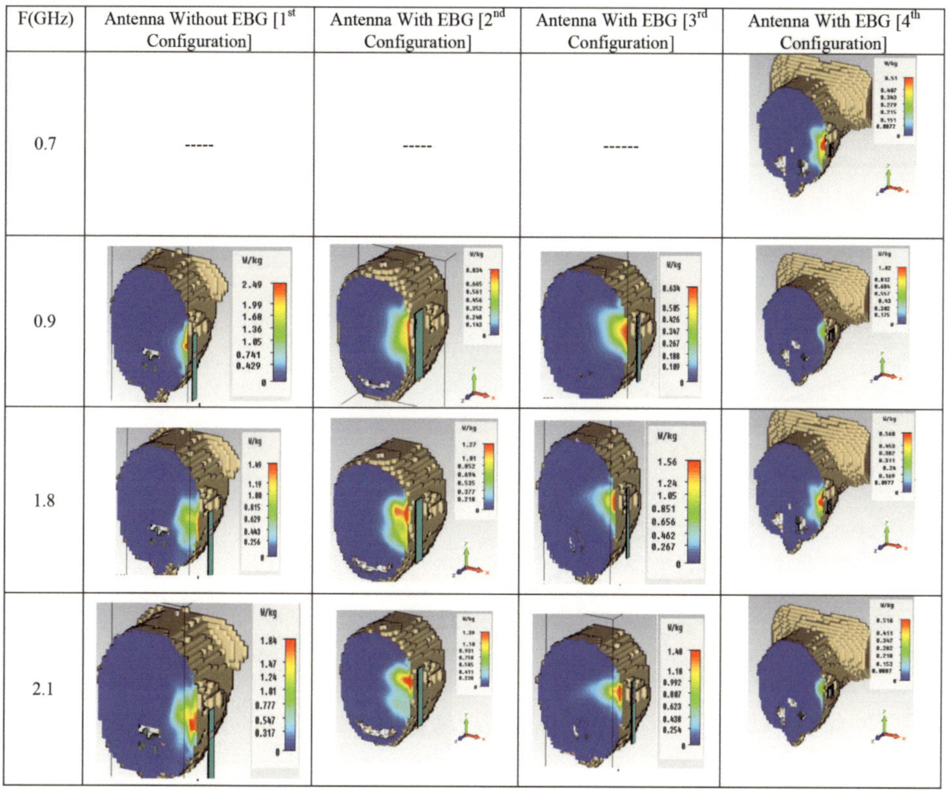

Figure 11. Simulated SAR distribution at 0.9, 1.8, and 2.1 GHz for the three antenna.

power level changes, the SAR values could be controlled by adjusting the separation between the antenna and the head as shown in **Figure 12**, which illustrates that the SAR values decrease with the increase of the separation distance between the antenna and the head. The SAR reduction factor (SRF) and the absorbed power are also tabulated in **Table 5**.

$$\mathrm{SRF} = 100\left(1 - \mathrm{SAR}_{\mathrm{EBG}}/\mathrm{SAR}_{\mathrm{No\ EBG}}\right)\% \tag{1}$$

Table 5 ensures that the antenna achieves the IEEE C95.1 and the international commission on non-Ionizing radiation protection (ICNIRP) standards [22]. The SAR value is inversely proportional to the distance between the antenna and the head as shown in **Figure 12**. Furthermore, the results are scalable when the power level changes.

Frequency (GHz)	First configuration	Second configuration	Third configuration	Fourth configuration	Absorbed power (rms) W	SRF (%) between (first and fourth)
0.7	–	–	–	0.536	0.0176	–
0.9	2.63 W/kg	0.872 W/kg	0.641 W/kg	1.03	0.03966	60.8
1.8	2.05 W/kg	1.27 W/kg	1.58 W/kg	0.788	0.086	61.5
2.1	1.78 W/kg	1.39 W/kg	1.56 W/kg	0.883	0.0971	50.39

Table 5. Maximum simulated SAR value (antenna closed to the human head).

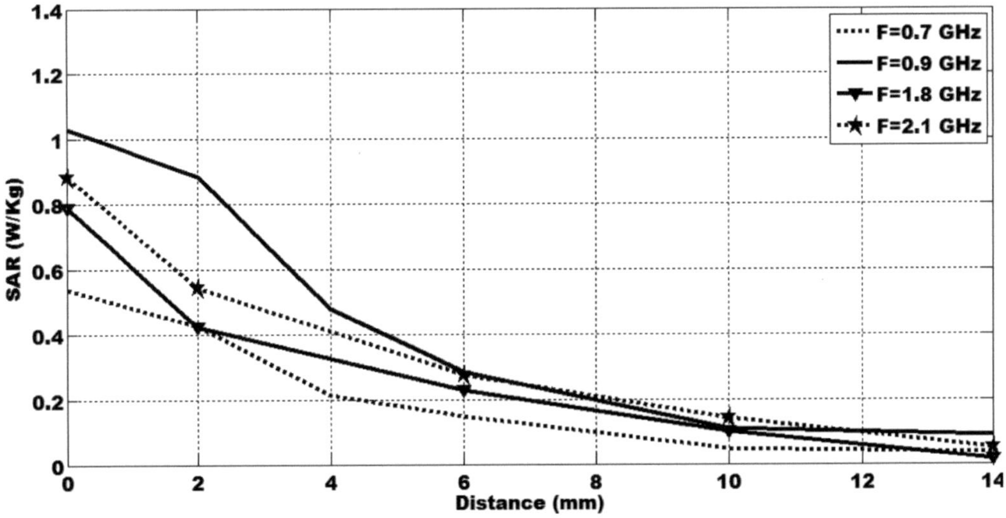

Figure 12. The variation of the SAR value at different frequencies against separation distance between head and antenna for 4th configuration.

4. Antenna in handset

There are several sorts of advanced mobile phones, each exhibiting different qualities and functions. One among these phones may be the "clamshell" alternately foldable portable phone, which has a large LCD, making it really attractive for the end client. It needs two fundamental printed circuit boards (upper and lower PCBs). In practice, the upper printed circuit board (PCB) includes LCD, camera, and speaker. The lower PCB on the other side typically includes radio frequency (RF) circuits and the battery of the phone, and the upper PCB includes keypads and buttons [4]. Another handset of these types is Samsung phone. In this section, we will introduce modeling of both the clamshell and Samsung phones.

4.1. Handset modeling

The advanced mobile phone has various components other than the PCB and the antenna. **Figure 13(a–c)** demonstrate the design of the cell phone without packaging for clamshell-type in the open state or talk state. The two grounds are of a similar size 40 × 85 mm^2. **Figure 13(d)** demonstrates the perspective of the cell phone in the close state or standby condition. In the open mode, the upper ground is slanted with respect to the focal line of the cell phone with an angle of 15°, which is acceptable inclination for the clamshell cell phone in the talk position. The upper ground is associated with the fundamental ground at a position far from the hinge position or the top edge of the main ground through an extended connecting strip of width 0.5 mm and length 8 mm. While in the close mode, the upper ground is parallel to the primary ground with a separation of 8 mm as appeared in **Figure 13(d)**. The antenna is amassed with the keypad model [20], battery, camera, speaker, RF circuit, and LCD. The installing zone of internal antenna has four choices. **Table 6** demonstrates the favorable position and weakness of every choice. "Impact of hand" implies that holding of client's hand will cause the move of the operating frequency, and transmitting and receiving become unstable. "Impact of head" defines the moving of operating frequency and raising SAR values. The length of link has an impact on the loss between the antenna and PCB. When a long link is utilized, the signal power will turn out to be low. **Table 6** demonstrates that the best position of the antenna is "Top of Key-Pad side."

The housing of the mobile is a casing of polyvinyl chloride material polyvinyl chloride (PVC) with permittivity of 2.8 and loss tangent of 0.019, where the total dimensions of the mobile in the closed state are 90 × 50 × 14 mm^3 and its wall thickness is 1 mm. **Figure 13(e)** and **(f)** show the mobile after housing with camera and speaker, the camera with 8.5 mm, and it is 6 mm thick. Opposite to the camera is a speaker with the dimensions of 20 mm length and 6 mm width. A large touch screen LCD with size of 70 × 40 × 2 mm^3 and a battery with volume 60 × 35 × 4 mm^3 are located parallel with a spacing of 1 mm and are connected to the main circuit board via connectors. **Table 7** shows the material properties of the mobile handset parts. **Figure 14** shows Samsung phone with standard dimensions 151.1 × 80.5 × 9.4) mm^3 [23], with the same housing material of clamshell mobile, speaker, camera, battery with size 90 × 60 × 4 mm^3, RF circuit, and large touch LCD with size 130 × 70 × 2 mm^3.

The return loss of the antenna in the two modes of the clamshell mobile (open mode and close mode) is shown in **Figure 15**. There is a slight shift between the results of return loss in the free

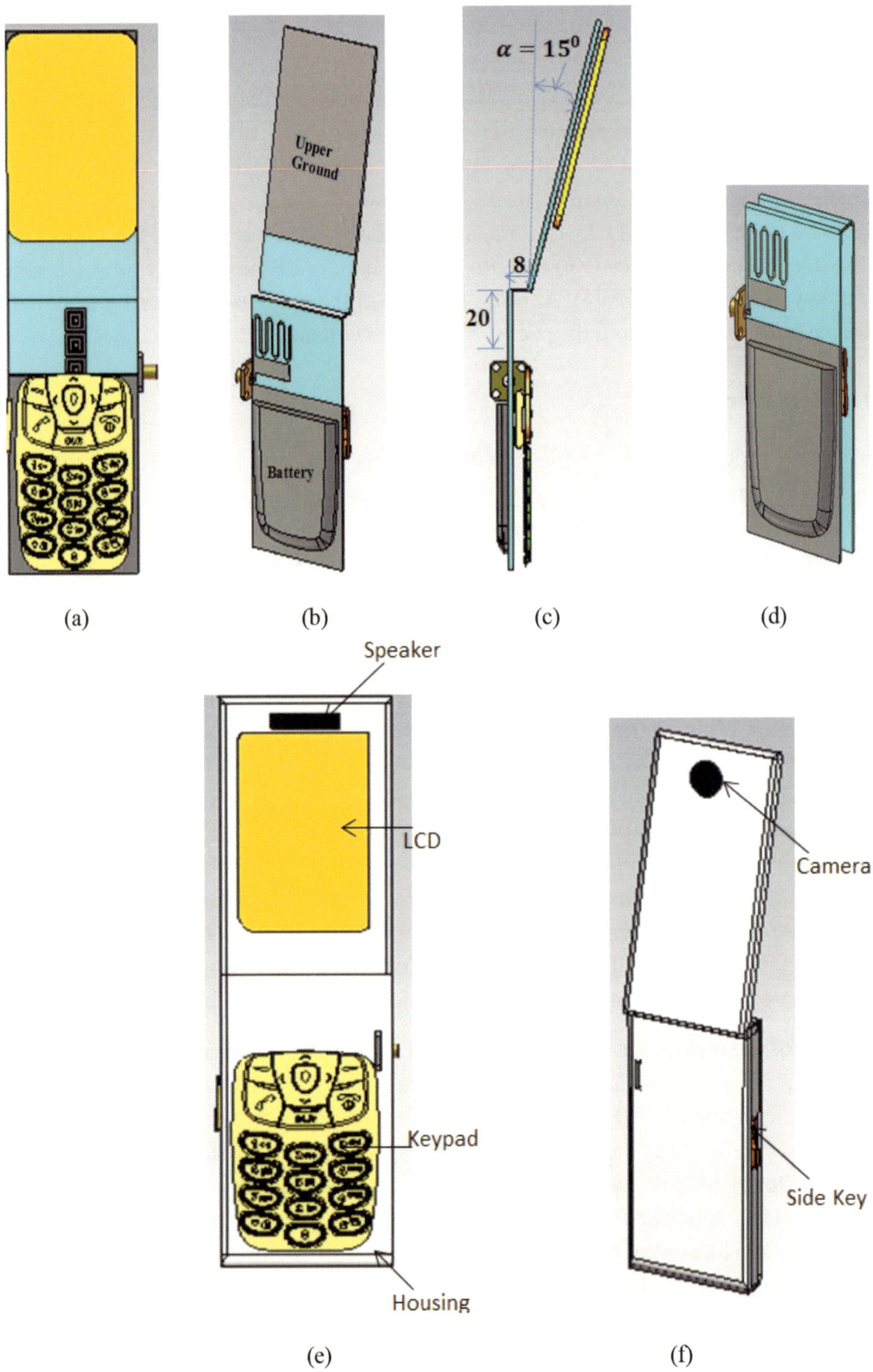

Figure 13. Geometry of the clamshell mobile. (a) Front view of clamshell without packaging. (b) Back view of clamshell without packaging. (c) Side view of clamshell without packaging. (d) Closed mode of clamshell without packaging. (e) Front view of clamshell with packaging. (f) Back view of clamshell with packaging.

Position	Influence for human body	Cable length
Top of LCD	Near the head, large SAR value	90 mm
Bottom of LCD	Little influence of head and hand	40 mm
Top of Key-Pad	Little influence of head and hand	0 mm
Bottom of Key-Pad	Great influence of hand	0 mm

Table 6. Comparison of antenna installing area.

space and in the clamshell handset. In the open case, the effect of the handset components is to mismatch and to shift the operating bands of the antenna, where the bandwidth decreases significantly in the low band. So, the dimensions of internal antenna are designed again to meet the effect of the handset components. The new dimensions of the antenna are L_2= 7 mm, L_4= 2.5 mm, and W_1= 18. In the close case, there is degradation in the impedance bandwidth over the bands. Nevertheless, the operating bandwidth is still better than 6:1 voltage standing wave ratio (VSWR), which is satisfactory to the clamshell phone in the closed case [4].

The requirements and regulations of the mobile handset are developed through the last years by the 3rd generation partnership project (3GPP). The test regulations on various standards were created or are progressing. The Cellular Telecommunications and Internet Association (CTIA)/The Wireless Association is a United States-based global association that serves the interests of the wireless industry by campaigning government organizations and helps with regulation settings. It is vital to test the last antenna execution that impacts the human body [24-32].

The absorption of human body can be characterized as head loss and hand loss when the mobile is in a talking or browsing position. The CTIA organization introduces four different body test cases for any mobile phone (free space test, browsing test, talking test, and talking with hand test) as shown in **Figure 16**. The return loss of those four cases is shown

Mobile parts		Material type	ε_r
LCD		LCD film	4.78
Battery		Perfect electric conductor (PEC)	–
RF circuit		PEC	–
Keypad	Digital keys, side key, function keys	Rubber	3.5
	Key lighter	LCD film	4.78
Camera		PEC	–
Speaker		PEC	–
Casing (housing)		PVC	2.8
PCB		FR4	4.5

Table 7. Characteristics of the mobile parts.

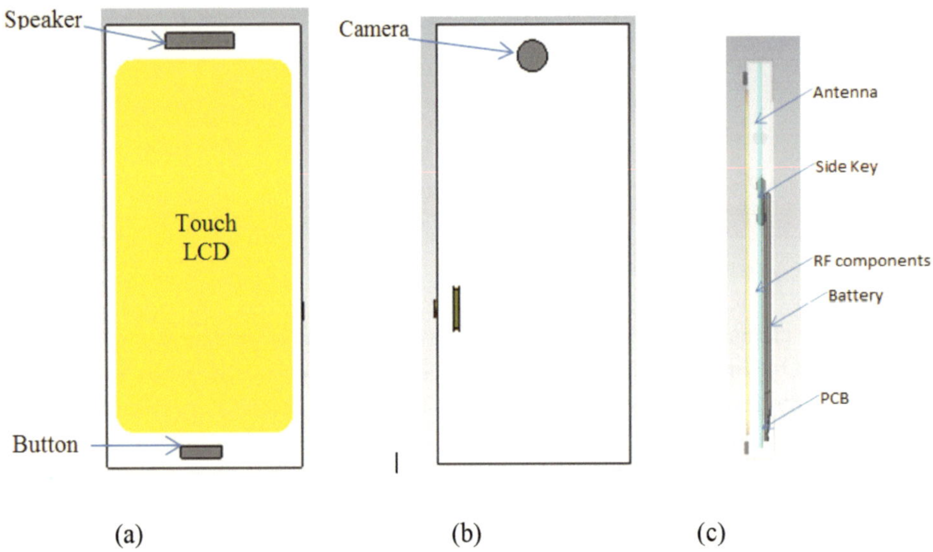

Figure 14. Structure of mobile elements. (a) Front view, (b) back view, and (c) side view.

in **Figure 17**. The hand and the head have little effect on the antenna impedance matching. Nevertheless, the impedance bandwidth over the operating bands is still acceptable for practical applications of the mobile phone.

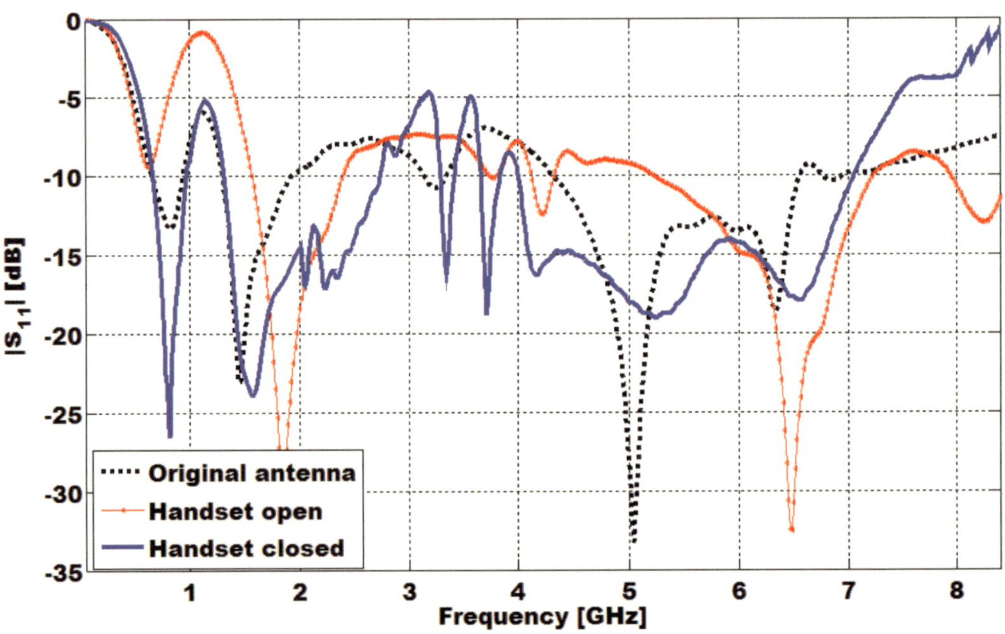

Figure 15. The simulated return loss of original antenna and antenna in handset.

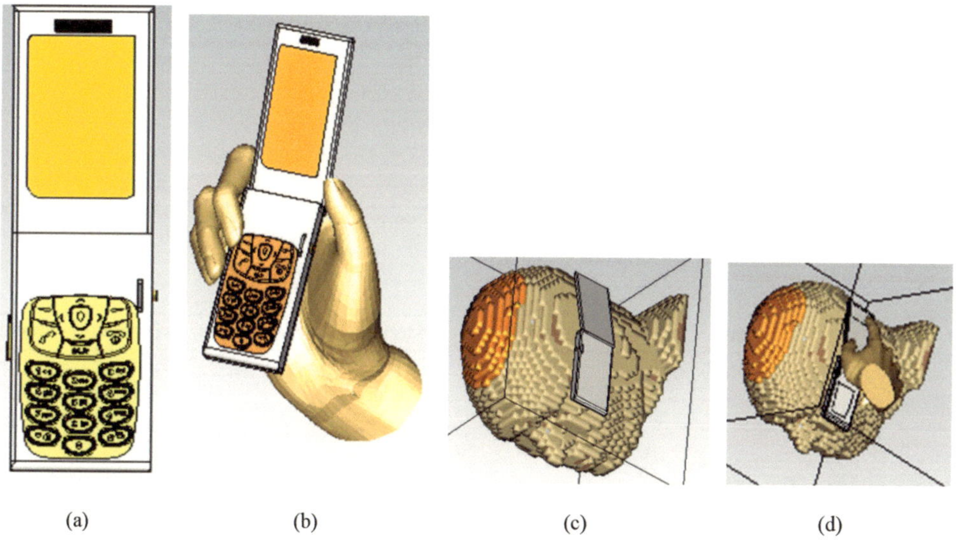

Figure 16. CTIA-defined four different test positions for clamshell mobile: (a) free space, (b) browsing mode, (c) talking position, and (d) talking position with hand.

A Hugo phantom is the specific anthropomorphic mannequin (SAM) phantom, which is defined for SAR measurement. The hand phantoms are defined for talking and browsing modes in evaluating the effects on different phone factors [26].

Table 8 shows the averaged 10 g SAR at the aforementioned operating frequencies for the original antenna and antenna in mobile handset when they are in close proximity to the body. The

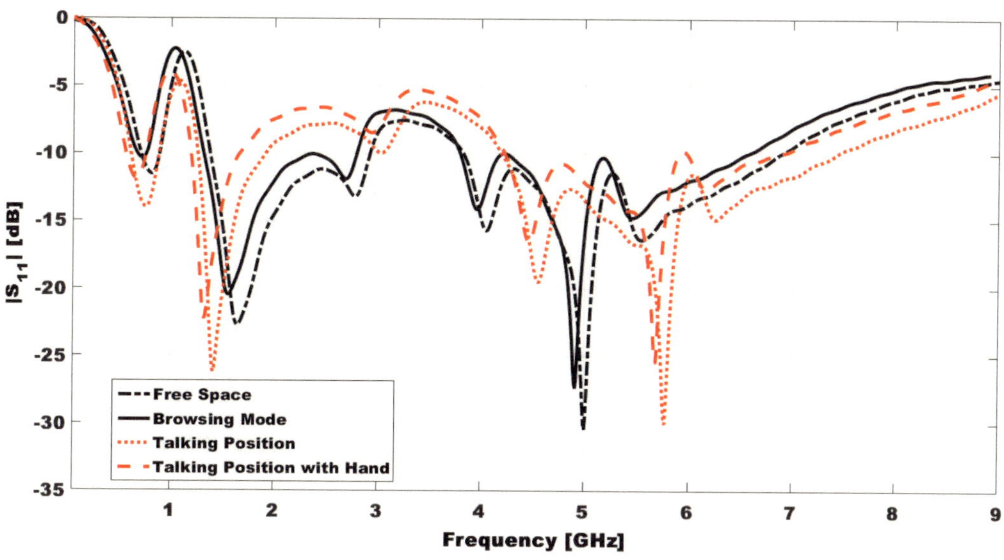

Figure 17. The simulated return loss of antenna in mobile handset in four different test positions.

F (GHz)	SAR (original antenna) W/kg	SAR (antenna in handset) W/kg	SAR (in car) W/kg
0.7	0.536	0.19	0.12
0.9	1.03	0.575	0.47
1.8	0.788	0.48	0.45
2.1	0.883	0.331	0.28

Table 8. Maximum SAR values of the proposed antenna.

F (GHz)	Original antenna (free space)	Handset (free space)	Handset antenna (with human model)
0.7	67%	62%	49%
0.9	65%	60%	61%
1.8	89%	92%	85%
2.1	83%	87%	79%

Table 9. Radiation efficiency.

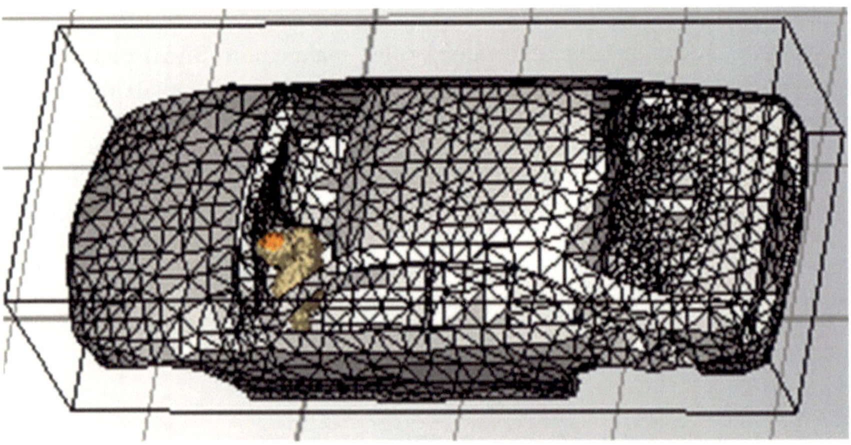

(a)

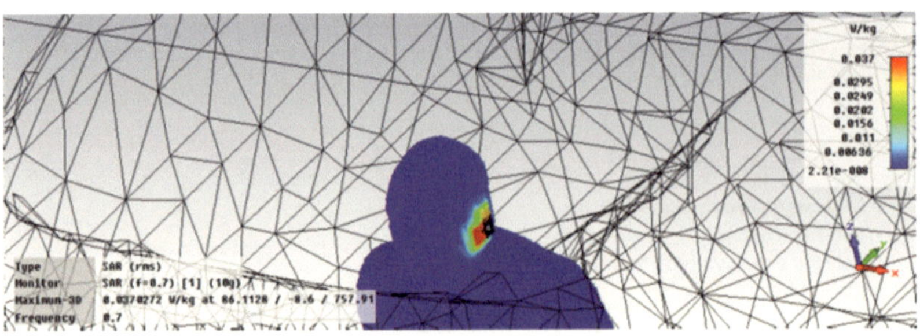

(b)

Figure 18. Human model and mobile handset placed within a car model. (a) Human inside car. (b) SAR distribution.

other important concern is the effect of the human body on the antenna performance as shown in **Figure 17**. **Table 9** shows the radiation efficiency of original antenna and antenna in handset in free space and talking position. We locate the human models inside the car, and position it 250 mm from the left and 350 mm from the top of the car model as shown in **Figure 18**. This, in turn, leads us to conclude that the electromagnetic environment of the car has very little effect on the estimated SAR values, and the SAR value is also tabulated in **Table 8**.

5. Conclusion

New four configurations of compact planar antennas that support the mobile applications, ISM band, and wireless services are introduced. The EBG structure is used to minimize the structure size of the antenna, give wide impedance BW, and reduce the SAR values. The SAR values of the all operating frequencies of the antenna are acceptable according to standards in four different modes (free space, talking mode, talking mode with hand, and browsing mode). Furthermore, the SAR values when the human body exists in a car satisfy the standards. Study of the effect of the human body on the antenna performance was also taken into considerations. The antennas have more compact size when compared to other antennas published in literatures. The antennas are simulated using the CST simulator and fabricated using the photolithographic technique. Very good agreement is obtained between the simulated and the experimental results.

Author details

Kamel Salah Sultan[1,2]*, Haythem Hussien Abdullah[1,3] and Esmat Abdel-Fatah Abdallah[1]

*Address all correspondence to: kamelsultan@eri.sci.eg

1 Electronics Research Institute, El-Tahreer, Giza, Egypt

2 Zewail City of Science and Technology, Giza, Egypt

3 Higher Institution of Engineering and Technology in Damietta new-NDETI, Egypt

References

[1] Sesia S, Toufik I, Baker M. LTE—The UMTS Long Term Evolution: From Theory to Practice. Chichester, UK: Wiley; 2009

[2] Johnston S. Review of ongoing research on radio frequencies and health 2002-05. In: Proc. 2nd. Intern. Workshop, Biological Effects of EMFs; Rhodes. 2002. pp. 1004-1013, Oct. 2002

[3] Villanueva RG, Aguilar HJ, Miranda RL. State of the art methods for low SAR antenna implementation. In: European Conf. on Antennas and Propag. (EuCAP);IEEE, Spain, pp. 1-4, 12-16 April, 2010

[4] Sultan KS, Abdullah HH, Abdallah EA, Hashish EA. Low SAR, miniaturized printed antenna for mobile, ISM, and WLAN services. IEEE, Antennas and Wireless Propagation Letters. 2013;**12**:1106-1109

[5] Sultan KS, Abdullah HH, Abdallah EA. Comprehensive study of printed antenna with the handset modeling. Microwave and Optical Technology Letters. 2016;**58**(4):974-980

[6] Sultan KS, Abdullah HH, Abdallah EA, Hashish E. Low SAR compact and multiband antenna for mobile and wireless communication. In: 2nd Middle East Conference on Antennas and Propagation, IEEE; Cairo, Egypt, pp. 1-5, 29-31 Dec., 2012

[7] Ang I, Guo YX, Chia YW. Compact internal quad-band antenna for mobile phones. Microwave and Optical Technology Letters. 2003;**38**(3):217-223

[8] Ciais P, Staraj R, Kossiavas G, Luxey C. Design of an internal quadband antenna for mobile phones. IEEE Microwave and Wireless Component Letters. 2004;**14**(4):148-150

[9] Tzortzakakis M, Langley RJ. Quad-band internal mobile phone antenna. IEEE Transcations on Antennas Propagation. 2007;**55**(7):2097-2103

[10] Ku CH, Liu HW, Lin SY. Folded dual-loop antenna for GSM/DCS/PCS/UMTS mobile handset applications. IEEE Antennas and Wireless Propagation Letters. 2010;**9**:998-1001

[11] Tang CL, Sze JY, Wu YF. A compact coupled-fed penta-band antenna for mobile phone application. In: Proceeding of Asia-Pacific Microwave Conference (APMC); 7-10 Dec. 2010; Yokohama, Japan. IEEE; 2010. pp. 2260-2263

[12] Li Y, Zhang Z, Zheng J, Feng Z, Iskander MF. A Compact hepta-band loop-inverted F reconfigurable antenna for mobile phone. IEEE Transactions on Antennas and Propagation. 2012;**60**(1):389-392

[13] Young CW, Jung YB, Jung CW. Octaband internal antenna for 4G mobile handset. IEEE Antennas and Wireless Propagation Letters. 2011;**10**:817-819

[14] Sultan KS, Abdullah HH, Abdallah EA, Hashish EA. Low SAR, planar monopole antenna with three branch lines for DVB, mobile, and WLAN. International Journal of Engineering and Technology IJET. 2014;**14**(1):70-74

[15] Sultan KS, Abdullah HH, Abdallah EA. Low SAR, simple printed compact multiband antenna for mobile and wireless communication applications. International Journal of Antenna and Propagation, Aug. 2014;**2014**:1-8

[16] Sultan KS, Abdullah HH. Multiband compact low SAR mobile hand held antenna. PIER Letter. 2014;**49**:65-71

[17] Yew-Siow Tay R, Balzano Q, Kuster N. Dipole configurations with strongly improved radiation efficiency for hand-held transceivers. IEEE Transactions on Antennas and Propagation. 1998;**46**(6):798-806

[18] Jung M, Lee B. SAR reduction for mobile phones based on analysis of EM absorbing material characteristics. Proceeding of Antennas and Propagation Society International Symposium; 22-27 June 2003; Columbus, OH, USA. 2003. Vol. 2, pp. 1017-1020

[19] Sultan KS, Abdullah HH, Abdallah EA, Hashish EA. Low SAR, compact and multiband antenna. In: Progress In Electromagnetics Research Symposium Proceedings, Taipei, Taiwan, pp. 748-751, 25-28 March, 2013

[20] CST Microwave Studio Suite 2011 User's Manual [Internet]. Available from: www.cst.com

[21] Gabriel S, Lau RW, Gabriel C. The dielectric properties of biological tissues II Measurements in the frequency range10 Hz to 20 GHz. Physics in Medicine & Biology. 1996;**41**:2251-2269

[22] IEEE C95.1-2005. IEEE Standards for Safety Levels with Respect to Human Exposure to Radio Frequency Electromagnetic Fields, 3 kHz to 300GHz. New York, NY: Institute of Electrical and Electronics Engineers; 2005

[23] http://www.gsmarena.com/samsung_galaxy_note_ii_n7100-4854.php

[24] Zhao K, Zhang S, Ying Z, He S. BSAR study of different MIMO antenna design for LTE application in smart mobile phone. In: Proc. IEEE Antennas Propag. Soc. Conf.; Chicago, IL. 2012

[25] Ying Z, Liu X. BSAR analysis of different antenna in different size of feature phone. Sony Ericsson Internal Report. 2010

[26] CTIA Certification Department Program. Test Plan for Mobile Station Over the Air PerformanceVMethod of Measurement for Radiated RF Power and Receiver Performance. [Internet]. Available from: www.ctia.org

[27] Federal Communications Commission (FCC), BSAR evaluation considerations for handsets with multiple transmitters and antennas, KDB 941255 D01, v01r05.2008

[28] Liu X, Ying Z. BBenchmark of feature mobile phone performance with different antennas including phone size, antenna size and body effects. Sony Ericsson Internal Report. 2010

[29] CTIA Test Plan for Wireless Device Over-the-Air Performance, Version 3.6.1, Nov. 2016 https://ctia.org/docs/default-source/certification/ctia-test-plan-for-wireless-device-over-the-air-performance-ver-3-6-1.pdf?sfvrsn=4

[30] 3GPP TS 34.114; User Equipment (UE)/Mobile Station (MS) Over-the-Air (OTA) Antenna Performance, V8.4.0, Jun. 2010

[31] Ying Z. Antennas in cellular phones for mobile communications. Proceedings of IEEE., July 2012;**100**(7);2286-2296

[32] International Non-Ionizing RaAssociation. Guidelines on limits on exposure to radio frequency electromagnetic fields in the frequency range from 100 kHz to 300 GHz. Health Physicdiation Committee of the International Radiation Protections. 1988;**54**(1):115-123